Edexcel GCSE (9–1)
Mathematics
Foundation

Practice, Reasoning and Problem-solving Book

Confidence • Fluency • Problem-solving • Reasoning

ALWAYS LEARNING

PEARSON

Published by Pearson Education Limited, 80 Strand, London WC2R 0RL.

www.pearsonschoolsandfecolleges.co.uk

Copies of official specifications for all Edexcel qualifications may be found on the website: www.edexcel.com

Text © Pearson Education Limited 2015
Edited by ProjectOne Publishing Solutions, Scotland
Typeset and illustrated by Tech-Set, Gateshead
Original illustrations © Pearson Education Limited 2015

The rights of Bola Abiloye, Gemma Batty, Phil Boor, Catherine Murphy and Claire Powis to be identified as authors of this work have been asserted by them in accordance with the Copyright, Designs and Patents Act 1988.

First published 2015

18 17
10 9 8 7 6 5 4 3 2

British Library Cataloguing in Publication Data
A catalogue record for this book is available from the British Library

ISBN 978 1 447 98359 0

Printed in Slovakia by Neografia

Acknowledgements
We would like to thank Glyn Payne for his work on this book.

The publisher would like to thank the following for their kind permission to reproduce their photographs:

Cover images: Front: Created by Fusako, Photography by NanaAkua

Every effort has been made to contact copyright holders of material reproduced in this book. Any omissions will be rectified in subsequent printings if notice is given to the publishers.

Contents

Welcome to Edexcel GCSE (9-1) Mathematics Foundation Practice, Reasoning and Problem-solving Book

This Practice Book is packed with extra practice on all the content of the Student Book – giving you more opportunities to practise answering simple questions as well as problem-solving and reasoning ones.

There is a section relating to every mastery lesson in the Student Book.

Icons alongside the questions show their level of difficulty. Questions in this book will range from 2 to 8.

The letters **P** and **R** are used to show where a question requires you to problem-solve or reason mathematically – essential skills for your GCSE.

Exam-style questions are included throughout to help you prepare for your GCSE exam.

QR codes link to worked examples from the Student Book which will help you with the question they are alongside.

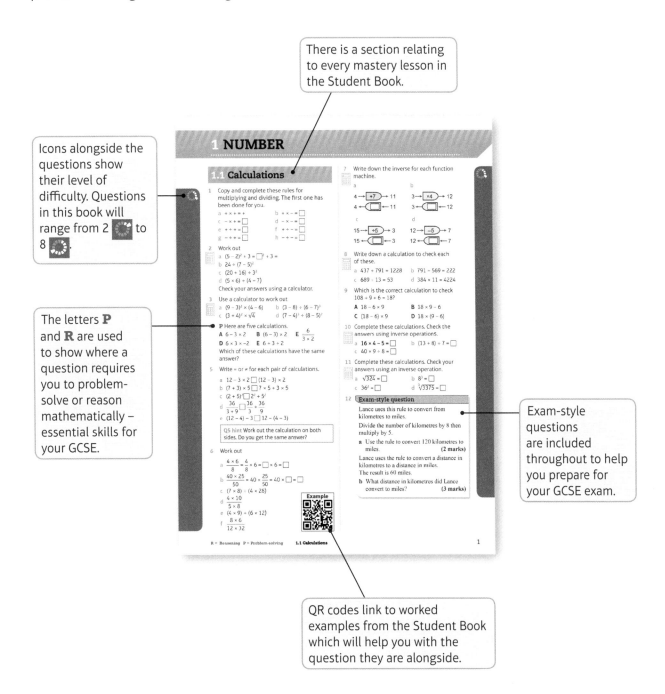

Problem-solving section

These are strategies that you have learned so far in the Student Book. They build up as you work through the book. Consider whether they could help you to answer some of the following questions.

A QR code is given in the problem-solving section where you learn a new strategy in that unit in the Student Book. Scan it to see the worked example and remind yourself of the strategy.

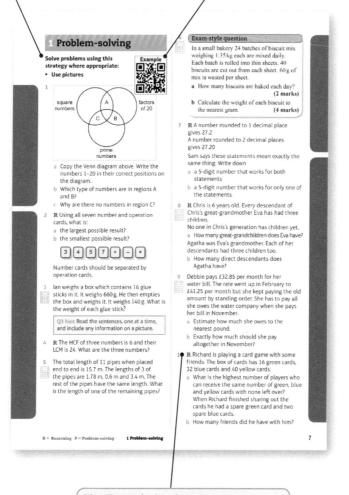

1 Problem-solving

Solve problems using this strategy where appropriate:
• Use pictures

Example

1
square numbers — factors of 20 — prime numbers Venn diagram with regions A, B, C

a Copy the Venn diagram above. Write the numbers 1–20 in their correct positions on the diagram.
b Which type of numbers are in regions A and B?
c Why are there no numbers in region C?

2 **R** Using all seven number and operation cards, what is:
a the largest possible result?
b the smallest possible result?

`3` `4` `5` `7` `+` `−` `×`

Number cards should be separated by operation cards.

3 Ian weighs a box which contains 16 glue sticks in it. It weighs 660 g. He then empties the box and weighs it. It weighs 140 g. What is the weight of each glue stick?

Q3 hint Read the sentences, one at a time, and include any information on a picture.

4 **R** The HCF of three numbers is 6 and their LCM is 24. What are the three numbers?

5 The total length of 11 pipes when placed end to end is 15.7 m. The lengths of 3 of the pipes are 1.78 m, 0.6 m and 3.4 m. The rest of the pipes have the same length. What is the length of one of the remaining pipes?

Exam-style question

In a small bakery 24 batches of biscuit mix weighing 1.35 kg each are mixed daily. Each batch is rolled into thin sheets. 40 biscuits are cut out from each sheet. 60 g of mix is wasted per sheet.

a How many biscuits are baked each day?
(2 marks)
b Calculate the weight of each biscuit to the nearest gram.
(4 marks)

7 **R** A number rounded to 1 decimal place gives 27.2
A number rounded to 2 decimal places gives 27.20
Sam says these statements mean exactly the same thing. Write down
a a 5-digit number that works for both statements
b a 5-digit number that works for only one of the statements.

8 **R** Chris is 6 years old. Every descendant of Chris's great-grandmother Eva has had three children.
No one in Chris's generation has children yet.
a How many great-grandchildren does Eva have?
Agatha was Eva's grandmother. Each of her descendants had three children too.
b How many direct descendants does Agatha have?

9 Debbie pays £32.85 per month for her water bill. The rate went up in February to £41.25 per month but she kept paying the old amount by standing order. She has to pay all she owes the water company when she pays her bill in November.
a Estimate how much she owes to the nearest pound.
b Exactly how much should she pay altogether in November?

10 **R** Richard is playing a card game with some friends. The box of cards has 16 green cards, 32 blue cards and 40 yellow cards.
a What is the highest number of players who can receive the same number of green, blue and yellow cards with none left over?
When Richard finished sharing out the cards he had a spare green card and two spare blue cards.
b How many friends did he have with him?

R = Reasoning P = Problem-solving **1 Problem-solving**

7

The **R** symbol indicates questions where you are required to reason mathematically. Questions in this section do not have the **P** symbol as they are all problem-solving.

1 NUMBER

1.1 Calculations

1 Copy and complete these rules for multiplying and dividing. The first one has been done for you.

a $+ \times + = +$ b $+ \times - = \square$

c $- \times + = \boxed{-}$ d $- \times - = \square$

e $+ \div + = \square$ f $+ \div - = \square$

g $- \div + = \square$ h $- \div - = \square$

2 Work out

a $(5 - 2)^2 + 3 = \square^2 + 3 =$

b $24 \div (7 - 5)^2$

c $(20 + 16) \div 3^2$

d $(5 \times 6) \div (4 - 7)$

Check your answers using a calculator.

3 Use a calculator to work out

a $(9 - 3)^2 \times (4 - 6)$ b $(3 - 8) \div (6 - 7)^3$

c $(3 + 4)^2 \times \sqrt{4}$ d $(7 - 4)^3 \div (8 - 5)^2$

4 **P** Here are five calculations.

A $6 - 3 \times 2$ **B** $(6 - 3) \times 2$ **C** $\dfrac{6}{3 \times 2}$

D $6 \times 3 \times -2$ **E** $6 \div 3 - 2$

Which of these calculations have the same answer?

5 Write = or ≠ for each pair of calculations.

a $12 - 3 \times 2 \square (12 - 3) \times 2$

b $(7 + 3) \times 5 \square 7 \times 5 + 3 \times 5$

c $(2 + 5)^2 \square 2^2 + 5^2$

d $\dfrac{36}{3 + 9} \square \dfrac{36}{3} + \dfrac{36}{9}$

e $(12 - 4) - 3 \square 12 - (4 - 3)$

> **Q5 hint** Work out the calculation on both sides. Do you get the same answer?

6 Work out

a $\dfrac{4 \times 6}{8} = \dfrac{4}{8} \times 6 = \square \times 6 = \square$

b $\dfrac{40 \times 25}{50} = 40 \times \dfrac{25}{50} = 40 \times \square = \square$

c $(7 \times 8) \div (4 \times 28)$

d $\dfrac{4 \times 10}{5 \times 8}$

e $(4 \times 9) \div (6 \times 12)$

f $\dfrac{8 \times 6}{12 \times 32}$

Example

7 Write down the inverse for each function machine.

a

$4 \rightarrow \boxed{+7} \rightarrow 11$

$4 \leftarrow \boxed{\square} \leftarrow 11$

b

$3 \rightarrow \boxed{\times 4} \rightarrow 12$

$3 \leftarrow \boxed{\square} \leftarrow 12$

c

$15 \rightarrow \boxed{\div 5} \rightarrow 3$

$15 \leftarrow \boxed{\square} \leftarrow 3$

d

$12 \rightarrow \boxed{-5} \rightarrow 7$

$12 \leftarrow \boxed{\square} \leftarrow 7$

8 Write down a calculation to check each of these.

a $437 + 791 = 1228$ b $791 - 569 = 222$

c $689 \div 13 = 53$ d $384 \times 11 = 4224$

9 Which is the correct calculation to check $108 \div 9 + 6 = 18$?

A $18 - 6 \times 9$ **B** $18 \times 9 - 6$

C $(18 - 6) \times 9$ **D** $18 \times (9 - 6)$

10 Complete these calculations. Check the answers using inverse operations.

a $16 \times 4 - 5 = \square$ b $(13 + 8) \div 7 = \square$

c $40 \times 9 \div 8 = \square$

11 Complete these calculations. Check your answers using an inverse operation.

a $\sqrt{324} = \square$ b $8^3 = \square$

c $36^2 = \square$ d $\sqrt[3]{3375} = \square$

12

> ### Exam-style question
>
> Lance uses this rule to convert from kilometres to miles.
>
> Divide the number of kilometres by 8 and then multiply by 5.
>
> **a** Use the rule to convert 120 kilometres to miles. **(2 marks)**
>
> Lance uses the rule to convert a distance in kilometres to a distance in miles. The result is 60 miles.
>
> **b** What distance in kilometres did Lance convert to miles? **(3 marks)**

1.2 Decimal numbers

1 Round these numbers to 1 decimal place.
   a 5.474   b 0.628   c 17.073   d 7.961

2 Round these numbers to 2 decimal places.
   a 8.0359   b 13.1829   c 0.1793   d 19.897

3 Round these numbers to 3 decimal places.
   a 7.4931   b 27.9056   c 8.1085

4 Round
   a 7.62 cm to the nearest mm
   b 8.7536 m to the nearest mm
   c 4.586 m to the nearest cm
   d 43.5364 km to the nearest m.

5 **P** Ayesha cuts a 2.4 m piece of ribbon into 9 equal pieces.
   What is the length of each piece of ribbon, correct to the nearest mm?

6 **P** Jared pays £9.60 for a multipack of chocolate buttons.
   There are 20 bags of chocolate buttons in the multipack. What is the cost per bag?

7 **P / R** Five friends buy rock concert tickets that cost a total of £229.
   How much is each ticket? Check your answer.

8 Work out
   a $0.3 \times 30 = 0.3 \times 3 \times 10 = 0.9 \times \square = \square$
   b 0.8×40
   c 0.4×600

9 Work out
   a 4×3 b 4×0.3
   c 0.4×3 d 0.4×0.3

10 Work out
   a 7.8×4 b 0.83×6
   c 92×0.4 d 173×2.31

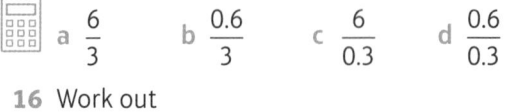
Example

11 **Exam-style question**

Sharon is buying drinks for a children's party.
Juice costs £1.15 per litre.
Cola costs £1.35 per litre.
Sharon buys 10 litres of juice and 5 litres of cola. She pays with a £20 note. Work out how much change she should get. **(3 marks)**

12 Fiona buys a box of 25 cupcakes.
   The cupcakes cost £2.15 each.
   How much does she pay?

13 Work out 0.6×0.06

14 **P** Katya multiplies two decimals and gets the answer 0.08
   What two decimals did she multiply? Is there more than one answer?

15 Work out
   a $\dfrac{6}{3}$ b $\dfrac{0.6}{3}$ c $\dfrac{6}{0.3}$ d $\dfrac{0.6}{0.3}$

16 Work out
   a $56.5 \div 5$ b $47.7 \div 3$ c $37.1 \div 7$

17 Work out
   a $5.68 \div 0.4$
   b $50.4 \div 0.12$
   c $560 \div 0.08$
   d $136 \div 1.6$

Example

> **Q17b hint** 0.12 has 2 decimal places, so multiply both numbers by 100 before dividing.

18 **R** Jake pays £4.32 for some table tennis balls.
   The balls cost 18p each.
   How many did he buy?

1.3 Place value

1 Write these numbers in figures.
   a 3.7 million   b 12.3 million
   c 7.36 million   d 1.625 million

2 Write these numbers in millions.
   a 27 800 000   b 8 750 000
   c 83 650 000   d 18 725 000

3 Work out these. Write your answers in figures.
   a 7.2 million + 4.65 million
   b 0.8 million + 9.15 million
   c 17.3 million − 4.68 million
   d 15.4 million − 0.85 million

4 Round
   a 638.729 to 4 s.f.   b 0.006 587 to 3 s.f.
   c 65 875 to 2 s.f.

> **Q4a hint** The 4th significant figure of 638.729 is 7. The next digit is 2, so leave the 7 as it is.

5 Round these to the number of significant figures shown.

   a 58 314 (3 s.f.)   b 0.0675 (1 s.f.)

   c 756 532 (2 s.f.)   d 56.386 (4 s.f.)

6 Work out

   a 600 × 800   b 3000 ÷ 600

   c 7000 × 400   d 80 000 ÷ 40

> **Q6a hint** 600 × 800 = 6 × 100 × 8 × 100
> = ☐ × 100 × 100
> = ☐

7 Estimate an answer for each calculation.

   a 384 × 625 ≈ 400 × ☐ =

   b $\dfrac{778}{38}$ c $\dfrac{18 \times 682}{985}$

   d $\dfrac{635 \times 3692}{525}$ e $\dfrac{185 \times 19.34}{0.48}$

   f $\dfrac{8.82 \times 9.41}{7.75}$ g $\dfrac{3.86 \times 8.75}{6.93}$

8 Check your answers to **Q7** using a calculator.

9 a **P** A is 15 million. B is 3 million. How many times larger is A than B?

In 2013, the population of Germany was 81.76 million. The population of Kazakhstan was 16.2 million.

   b Estimate the number of times greater the population of Germany was than the population of Kazakhstan.

10 **R** Use the information that 63 × 157 = 9891 to work out the value of

Example

   a 6.3 × 1.57

   b 98.91 ÷ 15.7

   c 9.891 ÷ 0.157

11 **R** Use the information that 287 × 29 = 8323 to work out the value of

   a 287 × 30   b 286 × 29

12 Calculate the value of $\dfrac{32.7 - 6.42}{15.6 - 4.53}$

   a Write down all the figures on your calculator display.

   b Give your answer correct to 2 decimal places.

13

> **Exam-style question**
>
> Calculate the value of
> $$\dfrac{15.38 \times (17.9 - 11.21)}{(35.8 + 16.3) \times 5.2}$$ **(3 marks)**
>
> a Write down all the figures on your calculator display.
>
> b Give your answer correct to 2 decimal places.

1.4 Factors and multiples

1 List the prime numbers between 30 and 50.

2 List the prime numbers between 70 and 90.

3 **R** The product of two prime numbers is always odd. Is this true or false?

4 List all the factors of 48.

5 List all the factors of

   a 24   b 42   c 50

6 Find the HCF of

   a 18 and 45   b 28 and 84

   c 32 and 80   d 12, 36 and 54

Example

7 a Write out the first 10 multiples of

    i 4   ii 5

   b Ring the common multiples of 4 and 5.

   c Which is the LCM of 4 and 5?

8 Find the LCM of

   a 3 and 8   b 6 and 15   c 4 and 7

   d 16 and 24   e 3, 5 and 9

9 Find the HCF and LCM of

   a 24 and 36   b 15 and 20

10 **R** Write two numbers with an HCF of 18.

11 **R** Write two numbers with an LCM of 60.

12 **P** Marta is making gift bags. She wants each to be the same with no items left over.

Marta has 32 loom bracelets and 48 chocolate bars. What is the greatest number of gift bags she can prepare?

> **Q12 hint** Find the HCF of 32 and 48.
> 32 = ☐ × 2   48 = ☐ × 3
> Use a diagram approach to check your answer.

13 **P** One set of Christmas lights flashes every 15 seconds. A second set flashes every 25 seconds. They both flash together at 8 pm exactly.
After how long will they next flash together?

14 ┌─────────────────────────────┐
Exam-style question

Sophie is making Christmas cards.

She needs a card and an envelope for each card.

There are 18 card blanks (for her to decorate) in a pack.

There are 30 envelopes in a pack.

Sophie bought exactly the same number of cards and envelopes.

a How many packs of cards and envelopes did she buy? **(3 marks)**

b How many cards can she make? **(2 marks)**
└─────────────────────────────┘

15 **P** Three cars are going around a rally race track. The first car completes a circuit every 0.25 hours, the second car every 0.4 hours and the third car every 0.5 hours. They start together at 2 pm. How long is it before all three cars are level again?

1.5 Squares, cubes and roots

1 Work out

a 4.7^2 b $\sqrt{54}$ to 1 d.p.

c 8.3^3 d $\sqrt[3]{127}$ to 3 s.f.

2 Work out

a 7^2 and $(-7)^2$ b 12^2 and $(-12)^2$

3 Use your answers to **Q2** to work out

a $\pm\sqrt{49}$ b $\pm\sqrt{144}$

┌─────────────────────────────┐
Q3 hint The symbol ± shows that you are being asked for the positive and the negative square root.
└─────────────────────────────┘

4 Work out

a 4^3 b $(-4)^3$

5 Use your answers to **Q4** to work out

a $\sqrt[3]{64}$ b $\sqrt[3]{-64}$

6 Work out

a $(-6.2)^2$ b $(4.3)^3$ to 3 s.f.

c $\sqrt{379}$ to 1 d.p.   d $\sqrt[3]{-749}$ to 3 s.f.

7 Find the positive square root of 5.76.

8 **P** A square blanket is 1.8 m by 1.8 m. What is the area of the blanket?

9 **P** The area of a square garden is 81 m². What is the length of one of the sides?

10 Find the pairs. Which is the odd card out?

1^4 16 125 2^3 5^3 8

3^4 3^3 1 81 32 27 2^5

11 Evaluate

a $3^2 \times 4$ b $4^2 + 5^2$ c $4^3 - \sqrt{36}$

d $3^3 - 6^2$ e $3^2 + 5^2 - 4^2$ f $2^3 \times 6 \times 5$

12 Work out

a $3.4^2 + 1.8^3$ b $7.6^3 - 8.4^2$

c $\sqrt{9^2 - 4.2^2}$ to 3 s.f.   d $\sqrt[3]{8.3 - 4.7}$ to 3 s.f.

13 **R** Between which two whole numbers does $\sqrt{56}$ lie?

┌──────────────┐
Example

└──────────────┘

14 **R** Work out $\sqrt{9} \times \sqrt{36}$ and $\sqrt{9 \times 36}$

What do you notice about your answers?

15 **R** $16 \times 36 = 576$

Use this fact to work out $\sqrt{576}$

16 Work out

a $\sqrt{2} \times \sqrt{2}$ b $\sqrt{7} \times \sqrt{7}$

What do you notice about your answers?

┌─────────────────────────────┐
Q16 hint $\sqrt{2} \times \sqrt{2} = \sqrt{2 \times 2}$
└─────────────────────────────┘

17 ┌─────────────────────────────┐
Exam-style question

Use your calculator to work out the value of $7.43 \times \sqrt{3}$.

a Write down all the figures on your calculator display. **(1 mark)**

b Write your answer to part **a** correct to 1 decimal place. **(1 mark)**
└─────────────────────────────┘

18

Exam-style question

Use a calculator to work out $\sqrt{\dfrac{32.6 \times 13.9}{7.4}}$

Write down all the figures on your calculator display. **(2 marks)**

19 Give your answer to **Q18** correct to 3 significant figures.

20 Work out $\sqrt{18}$. Give your answer

   a in surd form   b as a decimal.

21 Work out these. Give your answers in surd form.

   a $\sqrt{7^2 - 3^2}$ b $\sqrt{\dfrac{7 \times 12}{28 - 14}}$ c $\sqrt{5^2 + 10^2}$

   d $\sqrt{\dfrac{6 \times 8}{16}}$ e $\sqrt{12^2 - 6^2}$

1.6 Index notation

1 Copy and complete the pattern.

$2^1 = 2$

$2^2 = 2 \times 2 = \square$

$2^3 = 2 \times 2 \times 2 = \square$

$2^4 = 2 \times 2 \times 2 \times 2 = \square$

$2^5 = \ldots$

$2^6 = \ldots$

2 Copy and complete.

   a $6 \times 6 \times 6 \times 6 \times 6 \times 6 \times 6 = 6^{\square}$

   b $7 \times 7 \times 7 = 7^{\square}$

   c $8 \times 8 \times 8 \times 8 \times 8 = 8^{\square}$

3 Write as a product.

   a 3^6 b 4^5 c 9^4

4 Write each product using powers.

   a $5 \times 5 \times 5 \times 5 \times 5$

   b $8 \times 8 \times 8$

   c $3 \times 3 \times 3 \times 3 \times 3 \times 4 \times 4$

   d $5 \times 5 \times 7 \times 7 \times 7 \times 7$

5 Copy and complete using = or ≠.

   a $14^2 \square (2 \times 7)^2$

   b $15^2 \square 3^2 \times 5^2$

   c $3^3 \times 4^2 \square 3^2 \times 4^3$

   d $4^3 \times 4^2 \square 4 \times 4 \times 4 \times 4 \times 4 \times 4$

> **Q5 hint** Work out the calculation on both sides. Do you get the same answer?

6 Find pairs with the same value.

5^7 $5^3 \times 5^3$ $5^2 \times 5^3$

5^6 $5^2 \times 5^5$ 5^5

7 Write these expressions as a single power.

   a $6^3 \times 6^5$ b $7^7 \times 7^2$ c 8×8^5 d $9^3 \times 9^6$

8 a Work out $\dfrac{5 \times 5 \times 5 \times 5 \times 5 \times 5 \times 5 \times 5}{5 \times 5 \times 5 \times 5 \times 5}$ by cancelling.

   b Write your answer to part **a** as a power of 5.

   c Copy and complete.

$$\frac{5 \times 5 \times 5 \times 5 \times 5 \times 5 \times 5 \times 5}{5 \times 5 \times 5 \times 5 \times 5} = \frac{5^{\square}}{5^{\square}} = 5^{\square}$$

   d Copy and complete.

$$5^5 \div 5^2 = \frac{5^5}{5^2} = \frac{\square \times \square \times \square \times \square \times \square}{\square \times \square} = 5^{\square}$$

9 Write as a single power.

   a $6^8 \div 6^5$ b $7^5 \div 7$ c $8^6 \div 8^5$ d $4^7 \div 4^5$

10 Evaluate

   a $\dfrac{3^4 \times 3^2}{3^3}$ b $\dfrac{7^5 \times 7^4}{7^7}$ c $\dfrac{8^5 \times 8}{8^4}$

   d $\dfrac{4^4 \times 4^5 \times 4^6}{4^7}$

11 Write as a single power.

   a $(2^3)^2$ b $(5^3)^5$

   c $(4^5)^2$ d $(7^4)^4$

Example
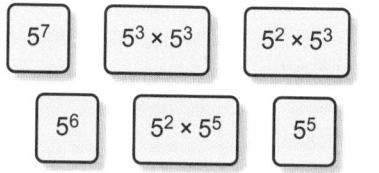

12

Exam-style question

Write as a power of 7

a $7^3 \times 7^2$ **(1 mark)**

b $7^8 \div 7^5$ **(1 mark)**

c $\dfrac{7^4 \times 7^5}{7}$ **(1 mark)**

d $(7^4)^2$ **(1 mark)**

13 Copy and complete the pattern.

$2^3 = 8$

$2^2 = 4$

$2^1 = 2$

$2^{\square} = 1$

$2^{-1} = \dfrac{1}{2}$

$2^{\square} = \dfrac{1}{2^2} = \dfrac{1}{4}$

$2^{\square} = \square = \square$

14 Write as a single power.

a $2^4 \times 2^{-1}$ b $2^5 \times 2^{-3}$ c $\dfrac{2^3}{2^5}$ d $\dfrac{2^4}{2^{-2}}$

15 Copy and complete.

a $1\,000\,000 = 10^{\square} = 1$ million

b $1\,000\,000\,000 = 10^{\square} = 1 \ldots$

c $1\,000\,000\,000\,000 = 10^{\square} = 1 \ldots$

16 Copy and complete the table of prefixes.

Prefix	Letter	Power	Number
tera	T	10^{12}	1 000 000 000 000
giga	G	10^9	
mega	M		1 000 000
kilo	k	10^3	
deci	d		0.1
centi	c	10^{-2}	
milli	m		0.001
micro	μ	10^{-6}	
nano	n		0.000 000 001
pico	p	10^{-12}	

17 Copy and complete.

a 1 centimetre (cm) = $\square$ m

b 1 gigatonne (Gt)= $\square$ kg

c 1 microgram (μg) = $\square$ g

d 1 nanosecond (ns) = $\square$ s

18 What is 10^9 tonnes in megatonnes?

1.7 Prime factors

1 a Complete these factor trees for 36.

Example

b Write 36 as a product of its prime factors.

2 Write these numbers as products of their prime factors.

a 28   b 48   c 60   d 96

3 **R** Farouk says, 'Prime numbers cannot be written as a product of prime factors.' Is this true or false? Show examples to explain.

4 Express these numbers as products of their prime factors. Draw Venn diagrams to find the HCF and LCM.

a 48 and 72   b 30 and 84

c 90 and 120

Example

5 What is 360 as a product of its prime factors?

A 40×9 **B** $5 \times 8 \times 9$ **C** $2^3 \times 3^2 \times 5$

D $2^3 \times 5 \times 9$ **E** $5 \times 2^3 \times 9$

6 $2^2 \times 3^2 \times 5$ is the prime factor decomposition of

A 45   **B** 90   **C** 120

D 180   **E** 360

7 **R** Use the information that $19 \times 23 = 437$ to find the LCM of

a 38 and 23   b 19 and 69

> **Q7 hint** What type of numbers are 19 and 23? What is the LCM of 19 and 23?
> $38 = 19 \times \square$ $69 = 23 \times \square$

8 **R** The number 600 can be written as a product of its prime factors.

$600 = 2^3 \times 3 \times 5^2$

Are these numbers factors of 600? Explain your answers.

a 24   b 90   c 120

9 **R / P** Lenny makes 336 plain biscuits, 216 chocolate biscuits and 144 nutty biscuits. Lenny wants to put all the biscuits into tins.

He wants to have the same number of biscuits in each tin.

The biscuits cannot be mixed.

He wants to buy the largest possible size of tin for the biscuits that can be filled leaving no spaces. How many biscuits should each tin hold?

10 **Exam-style question**

The number 3136 can be written as a product of its prime factors:

$$3136 = 2^6 \times 7^2$$

Write $\sqrt{3136}$ as a product of its prime factors. **(2 marks)**

1 Problem-solving

Solve problems using this strategy where appropriate:

Example
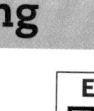

- **Use pictures.**

1 **R** Using all seven number and operation cards, what is
  a the largest possible result
  b the lowest possible result?

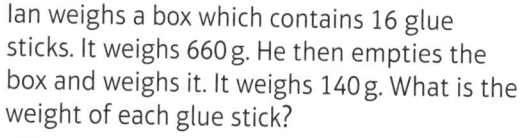

Number cards should be separated by operation cards.

2 Ian weighs a box which contains 16 glue sticks. It weighs 660 g. He then empties the box and weighs it. It weighs 140 g. What is the weight of each glue stick?

> **Q2 hint** Read the sentences, one at a time, and include any information on a picture.

3 The total length of 11 pipes when placed end to end is 15.7 m. The lengths of 3 of the pipes are 1.78 m, 0.6 m and 3.4 m. The rest of the pipes have the same length. What is the length of one of the remaining pipes?

4 **Exam-style question**

In a small bakery 24 batches of biscuit mix weighing 1.35 kg each are mixed daily. Each batch is rolled into one thin sheet. 40 biscuits are cut out from each sheet. 60 g of mix is wasted per sheet.

  a How many biscuits are baked each day? **(2 marks)**

  b Calculate the weight of each biscuit to the nearest gram. **(4 marks)**

5 **R** A number rounded to 1 decimal place gives 27.2
A number rounded to 2 decimal places gives 27.20
Sam says these statements mean exactly the same thing. Write down
  a a decimal number with 5 digits that works for both statements
  b a decimal number with 5 digits that works for only one of the statements.

6 Debbie pays £32.85 per month for her water bill. The rate goes up in February to £41.25 per month but she keeps paying the old amount by standing order. She has to pay all she owes the water company when she pays her bill in November.
  a Estimate how much she owes by October to the nearest pound.
  b Exactly how much should she pay altogether in November?

7
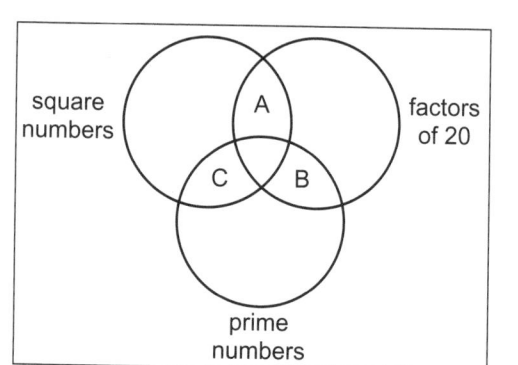

  a Copy the Venn diagram above. Write the numbers 1–20 in their correct positions on the diagram.
  b Which type of numbers are in regions A and B?
  c Why are there no numbers in region C?

8 **R** The HCF of three numbers is 6 and their LCM is 24. What are the three numbers?

9 **R** Richard is playing a card game with some friends. The box of cards has 16 green cards, 32 blue cards and 40 yellow cards.
  a What is the highest number of players who can receive the same number of green, blue and yellow cards with none left over?
When Richard finished sharing out the cards he had a spare green card and two spare blue cards.
  b How many friends did he have with him?

10 **R** Chris is 6 years old. Chris's great-grandmother Eva and each of her descendants has had three children.
No one in Chris's generation has children yet.
  a How many great-grandchildren does Eva have?
Agatha was Eva's grandmother. Each of her descendants had three children too.
  b How many direct descendants does Agatha have?

2 ALGEBRA

2.1 Algebraic expressions

1 Simplify by collecting like terms.

Example

a $5x + 3 + 4x + 5$

b $4p + 7q - 2p - q$

c $6d^2 + 3 - 4d^2 - 2$

d $5f^2 + 2f + f^2 - f$

> **Q1b hint** $1q = q$

2 **R** Isaac and Jake simplify $7x + 2y - x$.

Isaac's answer: $8x + 2y$

Jake's answer: $6x + 2y$

Who is correct?

3 a Add these terms together, moving clockwise around the loop.

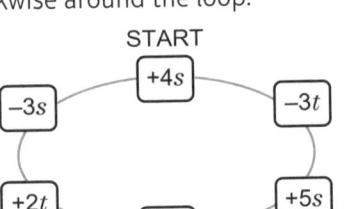

b Now add the terms together moving anticlockwise. Do you get the same result?

4 Simplify

Example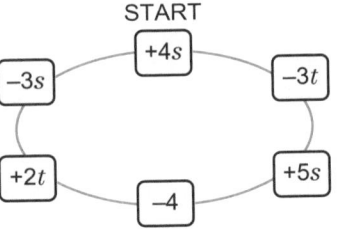

a $4 \times n$ b $g \times 15$

c $3m \times 4$ d $e \times f$

e $t \div 4$ f $d \div c$

5 **Exam-style question**

a Simplify $m + m + m - m + m$ **(1 mark)**

b Simplify $3g \times 5$ **(1 mark)**

c Simplify $5s + 4t + 2s + t$ **(2 marks)**

6 Write an expression for these.

a 12 more than m b 5 less than t

c 14 multiplied by n d 7 lots of x

e p divided by 4   f y halved

> **Q6a hint**
>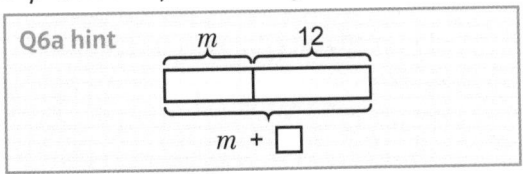
> $m + \square$

7 **R** Leili is x years old.

a Her sister is 3 years younger. Write an expression in x for Leili's sister's age.

b Her uncle is 4 times as old as Leili. Write an expression in x for Leili's uncle's age.

c Her brother is 2 years older than Leili. Write an expression in x for Leili's brother's age.

d Write and simplify an expression for the combined ages of all four.

8 Each pack of pens has x pens. Write an expression for the number of pens in

a 3 packs   b 5 packs

c 16 packs   d n packs

9 Write an expression for the perimeter P of the triangle.

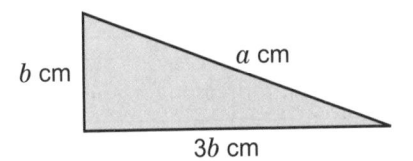

2.2 Simplifying expressions

1 Copy and complete.

a $6 \times 6 = 6^\square$ b $a \times a = a^\square$

c $7 \times 7 \times 7 = 7^\square$ d $t \times t \times t \times t = t^\square$

2 Copy and complete.

a $3^2 \times 3^3 = 3^\square$ b $y^2 \times y^3 = y^\square$

c $n^5 \times n^3 \times n = n^\square$

3 Write two terms that multiply together to give these answers.

a $\square \times \square = m^3$ b p^5 c q^{10}

4 Copy and complete.

a $6^5 \div 6^3 = 6^\square$ b $x^5 \div x^3 = x^\square$

c $7^{10} \div 7^3 = 7^\square$ d $y^{10} \div y^3 = y^\square$

5 Simplify

a $\dfrac{n^8}{n^3} = n^8 \div n^3 = \square$ b $\dfrac{e^9}{e^4}$

c $\dfrac{h^5}{h}$ d $\dfrac{x^{12}}{x^6}$

6 Simplify

a $2c \times 5d$

b $3m \times 8n$

c $5p \times 2q \times 6r$

d $7a \times b \times 4c$

e $-8x \times 5y$ f $b \times 5b$

g $7p \times 2q \times 3p$ h $6m \times -4m \times n$

Example

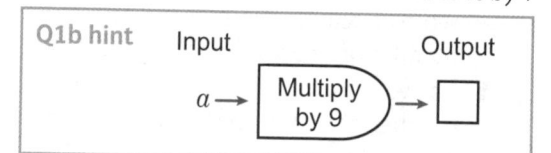

7 **R** Simon and Alexis work out the answer to $4c \times 5d \times 3e$

Simon's answer: $60cde$

Alexis's answer: $ecd60$

Whose answer is correctly written?

8 Find the product of each pair of expressions connected by the six lines.

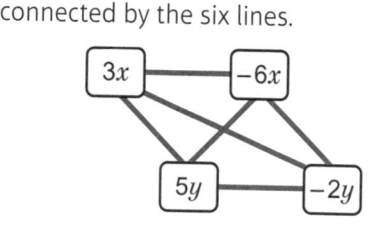

9 Simplify

a $\dfrac{12c}{4} = \dfrac{12}{4} \times c = \square \times c = \square c$

b $\dfrac{-36f}{4}$ c $\dfrac{45m}{15}$

d $\dfrac{3d}{9} = \dfrac{3}{9} \times d = \dfrac{\square}{\square} \times d = \dfrac{d}{\square}$

e $\dfrac{5x}{20}$ f $-\dfrac{9g}{3}$

g $\dfrac{9b}{15} = \dfrac{9}{15} \times b = \dfrac{\square}{\square} \times b = \dfrac{\square b}{\square}$

h $\dfrac{8s}{20}$ i $\dfrac{-8y}{-24}$

10 Alun says $n \div n = 0$. Janet says $n \div n = 1$. Who is correct?

11 Simplify

a $\dfrac{4n^3}{n} = 4 \times \dfrac{n^3}{n} = 4 \times \square =$

b $\dfrac{10x^5}{2x^3}$ c $\dfrac{3m^2}{12m^2}$

d $\dfrac{-12e^3}{4e^2}$ e $\dfrac{8r^4}{-32r^3}$

f $\dfrac{-16z^5}{-24z^2}$

12 **R** Insert the missing term in each equation. Choose from: $3m$ m^2 m

a $\dfrac{3m}{3} = \square$ b $\dfrac{\square}{m} = 3$

c $\dfrac{9m^2 n}{36\square} = \dfrac{n}{4}$ d $\dfrac{18m^2}{\square} = 6m$

e $\dfrac{6m^4 n}{6m^2 n} = \square$

2.3 **Substitution**

1 Write an expression for these statements. Use a to represent the starting number.

a Elena thinks of a number and subtracts 21

b Chris thinks of a number and multiplies it by 9

c Afzal thinks of a number and divides it by 7

Q1b hint

Input   Output

$a \rightarrow$ Multiply by 9 $\rightarrow \square$

2 Write an expression for these statements. Use n to represent the starting number.

a Sonali thinks of a number, multiplies it by 3 then adds 7

b Julia thinks of a number, multiplies it by 4 then subtracts 9

c Louis thinks of a number, doubles it then divides by 5

d Indira thinks of a number, divides it by 3 then adds 8

e Michael thinks of a number, subtracts 5 then divides by 3

3 Work out the value of these expressions when $p = 4$ and $q = 3$

Example

a $p + 2q$ b $q + 5$

c $pq - 3$ d $\dfrac{6p}{q}$

e $5p - 3q$ f $7pq - 34$

g $\dfrac{-27}{q}$ h $\dfrac{-32q}{p^2}$

4 Find the value of each expression when $x = 4$, $y = -3$ and $z = 5$

   a $8x + 8y$ **b** $2x + 3y$

   c $\dfrac{12x}{y}$ **d** $\dfrac{2 + x}{y}$

   e $3y^2$ **f** $x^2 + z^2$

   g $y^2 - z$ **h** $2y^2 + xz^2$

5

> **Exam-style question**
>
> $p = 9$
>
>   **a** Work out the value of $3p - 4$ **(2 marks)**
>
> $S = 4q + 9r$
>
> $q = -3$
>
> $r = 5$
>
>   **b** Work out the value of S.   **(2 marks)**

6 A bag of sweets has c chocolate sweets and t toffee sweets.

   a Write an expression in c and t for the total number of sweets in the bag.

   b Use your expression to work out the total number of sweets in the bag when $c = 8$ and $t = 11$.

7 **R** Mira makes b biscuits.

   a She gives one to her mum. Write an expression in b for the number of biscuits Mira has left.

   b She gives one third of the remaining biscuits to her brother. Write an expression in b for the number of biscuits Mira gives to her brother.

   c Use your answer to part **b** to work out how many biscuits Mira gives to her brother when $b = 10$.

8 **R a** A bag of apple slices costs m pence. Write an expression for the cost of 4 bags of apple slices.

   b A smoothie costs n pence. Write an expression for the cost of 3 smoothies.

   c Anya buys 4 bags of apple slices and 3 smoothies. Write an expression in m and n for the cost.

   d Use your answer to part **c** to work out how much Anya spends when $m = 45$ pence and $n = 95$ pence. Give your answer in pounds (£).

9 **P** A parcel delivery firm charges £15 for delivery of a parcel up to 10 kg in the UK, plus £2.50 for each kilogram over this amount.

   a The expression used to work out how much each parcel costs to send is £15 + £2.50k. What does k represent?

   b Copy and complete the table.

Total mass of parcel	Amount over 10 kg	$15 + 2.50k$
12 kg		
15 kg		
18 kg		
25 kg		

2.4 Formulae

1 **R** A normal size packet of crisps costs p pence.

   a Write an expression in p for the cost of a bumper size packet of crisps that costs 25 pence more.

   b Write a formula for the cost, C, of the bumper size packet.

2 **R** There are 24 segments in a chocolate orange.

   a Write an expression for the number of segments in c chocolate oranges.

   b Using your expression, write a formula for the total number n of segments in c chocolate oranges.

3 **R** A hair scrunchie costs s pence. Write formulae for the cost, C, of another hair scrunchie which costs

   a 25 pence less

   b twice as much

   c one third as much.

4 The formula to work out the density of a substance is $d = \dfrac{m}{V}$ where d = density (in grams per cm³, g/cm³), m = mass (in grams, g) and V = volume (in cm³).

Use the formula to work out the density of these substances where

   a mass = 120 g and volume = 40 cm³

   b mass = 80 g and volume = 16 cm³

5 A formula to calculate the surface area of a cube is $A = 6s^2$, where s is the side length of the cube.

Calculate the surface area of a cube when

a $s = 4\,cm$

b $s = 6\,cm$

c $s = 20\,cm$

6 The formula for the distance travelled during a journey is $d = st$ where d = distance, t = time and s = speed.

Work out the distance (d) in miles a car travels if it goes at

a 60 miles per hour for 3 hours

b 50 miles per hour for 4 hours

c 70 miles per hour for $2\frac{1}{2}$ hours

7 The formula for speed is $s = \dfrac{d}{t}$ where s = speed, d = distance and t = time.

Work out the speed, s, in kilometres per hour, km/h, when

a $d = 80\,km$ and $t = 2$ hours

b $d = 240\,km$ and $t = 8$ hours

c a cyclist travels a distance of 36 km in 3 hours

d a helicopter travels a distance of 320 km in 2 hours.

8 Exam-style question

$T = 6a - b$

$a = 4.5$

$b = 3.6$

Work out the value of T. **(2 marks)**

9 Which of these are formulae and which are expressions?

a $3x - 4$

b $d = st$

c $v = u + at$

d $\dfrac{d}{t}$

e $V = lwh$

f $v^2 + ma$

Q9 hint A formula always has an equals sign.

10 R Elaine is a waitress. She works h hours per day and is paid a rate of £p per hour.

Example

a Write a formula for Elaine's total pay per day, T, in terms of h and p.

b Use your formula to work out T when $h = 7$ and $p = 9$.

c At the end of each day Elaine and a colleague equally share the tips, t, they have earned together. Write a formula for R, the amount for tips that Elaine receives, in terms of t.

d Use your formula for part **c** to work out Elaine's share of the tips when $t = £50$.

e Use your answers to parts **b** and **d** to work out Elaine's total earnings for the day.

11 R A charm bracelet costs £20 for the bracelet and £12 for each charm.

a Write a formula for the total cost, T, of a bracelet with c charms.

b Use your formula to work out the cost of buying a bracelet with 8 charms.

12 In a right-angled triangle, $b^2 = c^2 - a^2$, where b is the length of the shortest side.

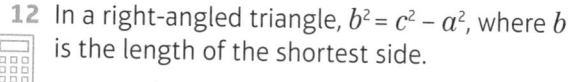
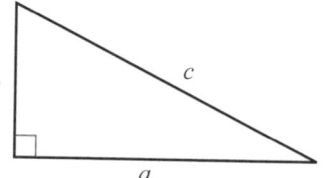

Find the length, b, of the shortest side, when

a $a = 8, c = 10$

b $a = 12, c = 13$

c $a = 4.7, c = 6.4$

d $a = 9.6, c = 11.5$

Round your answers to 1 d.p. where necessary.

2.5 Expanding brackets

1 Find the value of each expression when $p = 3$ and $q = 4$

a $3p(q - 2)$

b $7(p + 3q)$

c $(2p + q)^2$

d $q(4p - 7)$

e $(3q)^2$

2 Work out

a $5(6 + 3)$

b $5 \times 6 + 5 \times 3$

c $4(7 + 2)$

d $4 \times 7 + 4 \times 2$

3 Expand the brackets.

a $4(x + 3)$

b $3(m + 5)$

c $12(n - 3)$

d $7(4 - p)$

e $5(3t + 4)$

f $-3(b + 4)$

g $-6(4a + 1)$

h $-5(x - 3)$

i $-(n + 7)$

Example

Q3i hint $-(n + 7)$ is the same as $-1(n + 7)$

4 **R** Luke and Aliya expand $5(3m - 2n)$.
Luke's answer is $15m + 10n$.
Aliya's answer is $15m - 10n$.
Who is correct?

5 Expand
a $x(x + 2)$ b $m(m - 6)$ c $a(4a - 3)$

6 Write an expression for each statement.
Use a to represent the starting number.
a I think of a number, add 5 and multiply by 3.
b I think of a number, add 2 and multiply by 7.
c I think of a number, subtract 3 and multiply by 4.

7 **R** A CD holds $2t + 3$ tracks.
a Write and simplify an expression for the number of tracks on 4 CDs.
b When $t = 15$, how many tracks are there on 4 CDs?

8 Expand and collect like terms.
a $4(m + 3) + 5$
b $6(x - 4) + 12$
c $n(n - 5) + 3n$
d $-3(a + 4) + 20$
e $24 - 3(4 - 5f)$
f $13x - (2x + 5y)$

9 Expand and simplify
a $2(a + 3) + 3(a + 1)$
b $5(2x + 5) - 4(x + 3)$
c $3(4e - 3) + 2(e - 4)$

10 Work out the value of these expressions.
a $(4b)^2$ when $b = -3$
b $(3a - 2b)^2$ when $a = 8$ and $b = -3$
c $\dfrac{2a(2b - 5)^2}{ab}$ when $a = -5$ and $b = 2$
d $\dfrac{(14a + 10b)^2}{b^2}$ when $a = 6$ and $b = -8$

11 **R** Gill is y years old and her brother is 4 years older. Her cousin is half the age of Gill's brother.
a Write an expression for the age of Gill's brother.
b Write an expression for the age of Gill's cousin.
c Use your answer to part **b** to work out the age of Gill's cousin when Gill is 18.

12 **R** An electrician charges a £25 callout fee plus £30 per hour for each job she does.
a A particular job takes j hours. Write an expression in j for the cost of the job.
b Write a formula for the cost, C, in terms of j.
c The electrician does four identical jobs, that each take j hours. Write a formula in j for her earnings, E.
d When j is 5, how much does the electrician earn for the four jobs?

13 **R** A joint of meat needs to be cooked for 60 minutes per kilogram plus 30 minutes extra. The oven needs to be warmed for 20 minutes before the meat can be cooked. The expression $20 + 60 (w + 0.5)$ represents the total amount of time the oven is on, in minutes.
a Write what the terms in the expression represent.
b $w = 2$. How long was the oven on for (in minutes)?

2.6 Factorising

1 Find the HCF of
a −15 and 10   b 12 and −8

Q1a hint Factors of −15 are −15, −5, −3, −1, 1, 3, 5, 15

2 Write the missing terms.
a $3(s + \Box) = 3s + 12t$
b $\Box(s - 6) = 6s - 36$
c $5s(\Box + \Box) = 25s^2 + 5st$
d $\Box(3s - 1) = 6st - 2t$

3 a Copy and complete to find the highest common factor (HCF) of $12n$ and 30.
i The factor pairs for 30 are 1×30, $2 \times \Box$, $\Box \times 10$, $\Box \times \Box$
ii The factor pairs for $12n$ are $1 \times 12n$, $\Box \times 6n$, $3 \times \Box$, $4 \times \Box$, $6 \times \Box$, $12 \times \Box$
iii The HCF is $\Box$
b What is the HCF of $36x$ and 54?
i Work out the factor pairs for each term.
ii By looking at all these factors, work out the HCF.
c Find the HCF of
i $8m$ and 12   ii $24y$ and 18

4 Copy and complete. Check your answers by expanding the brackets.

Example

a $7x + 28 = 7(x + \square)$
b $12p - 18 = 6(\square - 3)$
c $15n + 9 = 3(\square + 3)$
d $12s - 27 = \square(4s - \square)$

5 **R** Julia and Helen factorise $12d + 18$
Julia says the answer is $6(2d + 3)$.
Helen says the answer is $3(4d + 6)$.

Who is correct?

6 Factorise completely

a $8e + 24$
b $4m - 20$
c $18p + 9$
d $25 - 15t$

7 **R** Karl and Rashid factorise $18m - 6$
Karl's answer is $6(3m - 0)$.
Rashid's answer is $6(3m - 1)$.
Who is correct? What mistake has the other student made?

8 Work out the HCFs of

a ab and a
b m^2 and $2m$
c pq and pr
d $8ef$ and $24f$
e $7d^2$ and $-21d$
f $8ab$ and $12b^2$

9 Factorise

Example

a $e^2 + 2e$
b $5d^2 + 6d$
c $3pq + 6q$
d $6ef - 3f$

10 Choose the correct factorisation for each expression.

a $2a^2 + 5a$ **A** $\boxed{a(2a + 5)}$ **B** $\boxed{a(2a + 5a)}$

b $6ab - 7b$ **A** $\boxed{b(6a - 7b)}$ **B** $\boxed{b(6a - 7)}$

c $b^2 - b^3$ **A** $\boxed{b(1 - b^2)}$ **B** $\boxed{b^2(1 - b)}$

d $7a - a^2$ **A** $\boxed{a(7 - a)}$ **B** $\boxed{7a(1 - a)}$

11 Factorise these completely. Expand your answers to check them.

a $x^2 + 7x$
b $6x^2 - 24x$
c $8x + 12y$
d $12y^2 - 3y$

12 Exam-style question

  a Factorise $6a + 14b$ **(1 mark)**

  b Factorise $a^2 + 9a$ **(1 mark)**

13 Which of these are true for *all* values of b rather than just *some* values of b?
Rewrite any identities that you find, replacing = with $\equiv$

a $b + 4 = 6$
b $3b + 9 = 3(b + 3)$
c $b^2 = 9b$
d $4b + 11 = 11 + 4b$

14 Use $\equiv$ to write an identity for each of these expressions.

a $3x + 2x$
b $\frac{1}{3}x$
c $2(x - 3)$
d $x + 4$

15 State whether each of these is an expression, formula or identity.

a $s = \dfrac{d}{t}$
b $x^2 - 3x = x(x - 3)$
c $x^2 - 3x + 4$
d $9x^2 = (3x)^2$
e $s^2 + \frac{1}{2}as$
f $V = IR$

16 Write $\neq$ or $\equiv$ in each box.

a $2.5x \square \frac{1}{2}(5x + 1)$

b $5(x - 1) \square 5x$

c $12a + 6 \square 6(2a + 1)$

d $3x - x^2 \square -x(x - 3)$

2.7 Using expressions and formulae

1 Write a formula for P, the perimeter of each shape.

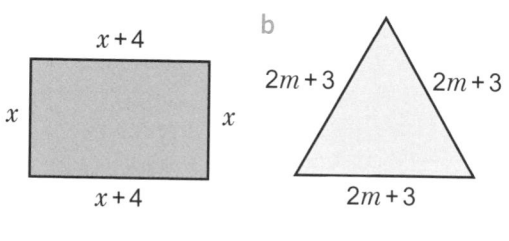

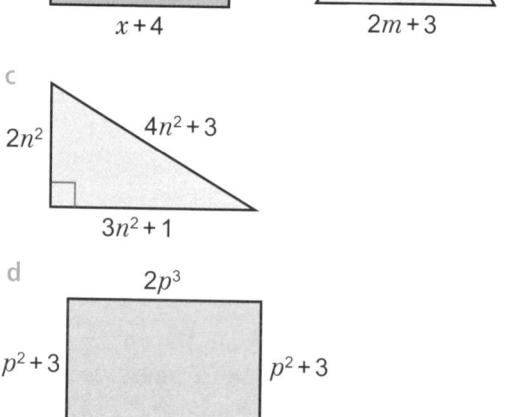

2 Write an expression using u as the unknown starting number.
  a I think of a number, multiply it by 4 then add 3.
  b I think of a number, subtract 3 and then multiply by 5.
  c I think of a number, multiply it by 4 and divide by 3.
  d I think of a number, multiply it by itself then multiply by 2.
  e I think of a number, square it, then multiply by the original number again.
  f I think of a number, square root it and then add 2.

> **Q2 hint** Use a function machine.

3 **R** Crisps are sold in multipacks or boxes. A multipack contains p packets of crisps and a box contains q packets of crisps.
  a Write an expression for the number of packets of crisps in 4 multipacks.
  b Write an expression for the number of packets of crisps in 5 boxes.
  c Grace buys 4 multipacks and 5 boxes for a tuck shop.
    Write an expression in terms of p and q for the number of packets of crisps she buys.
  d Grace buys m multipacks and b boxes of crisps.
    Write an expression in terms of p, q, m and b for the number of packets of crisps she buys.
  e When $m = 5$ and $b = 3$, use your answer to part **d** to work out how many packets of crisps Grace buys.

4 **R** Darren scores m marks in his French listening exam.
  a Write an expression in terms of m for
    i his French speaking marks, 10 marks less than for his listening exam
    ii his French reading marks, twice as much as the marks for his speaking exam
    iii his French writing marks, twice as much as for his listening exam, plus another 6 marks.
  b Write and simplify an expression for Darren's total marks in his French exam.
  c Darren scored 30 marks in his French listening exam. Use your answer to part **b** to work out his total marks in his French exam.

5 The formula to work out the voltage, V, of an electrical circuit is $V = IR$ where I = current and R = resistance.
  Work out V, in volts, when $I = 5$ amps and $R = 12$ ohms.

6 Use the formula $E = \frac{1}{2}mv^2$ to work out E (in joules, J) when
  a $m = 4$ kg, $v = 5$ m/s
  b $m = 6$ kg, $v = 10$ m/s

> **Example**

7 The formula $a = \dfrac{v - u}{t}$ gives the acceleration of an object where a = acceleration (m/s²), u = initial velocity (m/s), v = final velocity (m/s) and t = time (s).
  Work out the value of a when
  a $u = 15$ m/s, $v = 60$ m/s, $t = 5$ s
  b $u = 10$ m/s, $v = 46$ m/s, $t = 6$ s

8 Use the formula $u^2 = v^2 - 2as$ to work out the value of u when $v = 0$ m/s, $a = -4$ m/s² and $s = 8$ s

9 **Exam-style question**

The selling price, S, of an item in a shop can be worked out using the formula $S = \dfrac{7C}{5} + 4$ where C is the cost price.
  a Work out S when $C = £15$. **(2 marks)**
  b Which is cheaper to buy, an item with a selling price of £32 or an item with a cost price of £22?
    Give reasons for your answer. **(2 marks)**

10 **P** To cook a joint of lamb takes 20 minutes per kg plus an extra 30 minutes.
  a How long does it take to cook a 3.5 kg joint of lamb?
  b Write a formula for the number of minutes, M, it takes to cook a joint of lamb that weighs w kg.
  c Use your formula to find M when $w = 5$.
  d What time should you put a 2.5 kg joint of lamb in the oven to be ready for 6 pm?

2 Problem-solving

Solve problems using these strategies where appropriate:
- **Use pictures**
- **Use smaller numbers.**

1 **R** a Using $p = 5$, show that $2p^2$ is not the same as $(2p)^2$.
  b $2p^2$ can be the same as $(2p)^2$. What does p have to be for this to be true?

2

Exam-style question

Pat has x cards.

Jim has 4 more cards than Pat.

a Write down an expression, in terms of x, for the number of cards Jim has. **(1 mark)**

Lex has 2 times as many cards as Pat.

b Write down an expression, in terms of x, for the number of cards Lex has. **(1 mark)**

3 **P** A factory makes nails. In one day, it makes 26 280 nails. Nine conveyors put the nails into boxes, each containing 40 nails.

a How many boxes does each conveyor produce in one day?

b Write an expression for the number of boxes produced by each conveyor when the factory makes n nails and puts them into boxes containing b nails using c conveyors.

Q3 hint Read each sentence, one at a time. Replace each number with a smaller number and draw a picture to represent it.

4 **R** Andy and Liam are going out for the day. Their mum gave Andy y pence. She gave Liam 50p more. Their dad gave Andy 70p and Liam y pence.

a Write an expression for the difference between their amounts of money.

b Can you tell who has more from the information given? Explain your answer.

c How much does y have to be for them to have the same amount of money?

5 A spring is s cm long. It stretches e cm for each 50 g weight attached to it.

a How long does the spring get when 200 g is attached to it?

b How long does it get when w g is attached to it?

6 To calculate the final velocity (v) of an object, you multiply the acceleration (a) by the time taken (t) and then add the initial velocity (u).

a Write the formula for calculating final velocity (v) in terms of a, t and u.

b Use your formula to work out v when $u = 13$ m/s, $a = -2$ m/s² and $t = 5$ s.

7 Henry has p friends. They each give him £b to buy a ball for the group. He gives them £k change each.

a Write an expression in terms of b, k and p for the cost of the ball.

b Use your expression to calculate the cost of the ball if there are 8 friends who each give him £5 and and he gives them £2.24 change each.

8 A farmer is enclosing a square field with area 318 096 m². He wants to put the fence posts 2 m apart.

a How many fence posts will he need?

b The fencing comes in rolls of 120 m. How many rolls does he need?

c i Write an expression for the number of fence posts he needs for a field with area A m².

ii How will your expression change if the farmer decides to put the fence posts 3 m apart?

d Write an expression for number of rolls of fencing he needs for a field with area A m², if the fencing comes in rolls of f m.

9 Seema is knitting a scarf. In each row, she knits 5 plain stitches then 3 purl stitches, 5 plain then 3 purl, and so on. Each row has this pattern 4 times.

a She knits 17 rows. How many stitches is that?

b How many stitches in a rows?

Different patterns can be made by varying the numbers of plain and purl stitches. Use d for the number of plain stitches, e for the number of purl and r for the number of times that the pattern is repeated in each row.

c Write a formula for n, the number of stitches in a rows.

d Put the values for Seema's scarf into your formula from part **c**. Do you get the same answer as in part **a**?

10 **R** Two expressions have a highest common factor of a and a lowest common multiple of a^2b.

What could the expressions be?

3 GRAPHS, TABLES AND CHARTS

3.1 Frequency tables

1 A primary school teacher records the number of pupils in each class.
She records the data in a table.

Number of pupils in a class	Number of classes
28	4
29	1
30	3
31	2

   a Work out how many classes there are in the school.

   b Work out the total number of pupils in the school.

2 A swimming pool records the ages of people at an 11–18s' evening.
12, 15, 17, 13, 14, 16, 16, 11, 11, 13, 15, 16, 15, 16, 17
Copy and complete the grouped frequency table.

Age (years)	Tally	Frequency
11–12		
13–14		

3 For each inequality, list the numbers in the cloud that belong to it.

   a $3 < x \leqslant 10$ b $3 \leqslant x < 9$

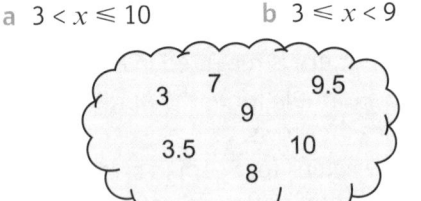

4 A Pilates instructor records the ages of people in the class.
27, 32, 34, 25, 37, 42, 49, 17, 19, 25, 52, 18, 22, 41, 54, 45, 44, 38, 19, 22

   a Copy and complete the grouped frequency table for this data.

   b What is the least common age group?

   c How many people are aged 25 or more?

Age, a (years)	Tally	Frequency
$15 \leqslant a < 25$		
$25 \leqslant a < 35$		
$35 \leqslant a < 45$		
$45 \leqslant a < 55$		

5 Alice measured the heights of 15 seedlings to the nearest mm.
15, 17, 19, 13, 14, 16, 17, 21, 17, 18, 19, 18, 19, 12, 17

   a Is this data discrete or continuous?

   b Design and complete a suitable grouped frequency table.

Example

6 Abene asked some students which was their favourite type of 'fast food' and recorded her results.
burger, pizza, burger, fish and chips, burger, fried chicken, fried chicken, pizza, burger, burger, pizza, falafel, burger, fried chicken, fried chicken, burger, pizza, burger, fried chicken
Design and complete a data collection sheet for Abene's data.

7 Jacob records the number of cars passing his house each minute.
11, 14, 11, 9, 7, 11, 10, 8, 12, 15, 18, 15, 14, 18, 16
Design and complete a data collection sheet for Jacob's data.

8 An athlete records the time it takes him to run the 100 m.
10.2, 10.1, 10.0, 10.9, 10.9, 10.0, 10.1, 10.5, 10.2, 10.7, 10.8, 10.0
Design and complete a data collection sheet for this data.

> **Q8 hint** Group the data.

3.2 Two-way tables

1 Here is part of a train timetable.

Petersfield	1031	1048	1117	1153
Haslemere	1046	1103	—	—
Guildford	1117	1134	1201	—
London Waterloo	1149	1206	1233	1259

a How long does the first train from Petersfield take to travel to London Waterloo?

b Anika wants to arrive in London before 1245. What is the latest train she can catch from Petersfield?

c Do all the trains take the same time from Petersfield to Guildford? Explain how you know.

2 The distance chart shows distances in miles between four airports.

Heathrow, London			
188	Charles de Gaulle, Paris		
397	297	Frankfurt	
891	685	601	Fiumicino, Rome

How far is it from

a London Heathrow to Frankfurt

b Charles de Gaulle, Paris, to Fiumicino, Rome

c Frankfurt to Fiumicino, Rome?

3

Exam-style question

The chart shows the time, in minutes, it takes Mrs Annery to walk between villages around Petersfield.

Petersfield				
17	Stroud			
15	15	Steep		
12	25	8	Sheet	
47	31	35	52	East Meon

a Write down the time it takes her to walk between Steep and East Meon. **(1 mark)**

Mrs Annery walks from Stroud to East Meon. 17 minutes into the journey she stops for a rest.

b Work out how much longer she must walk to reach East Meon. **(2 marks)**

c Write down the names of the two villages that Mrs Annery can get between the most quickly. **(1 mark)**

4 The table shows the colour and size of hoodies that were ordered for students in Year 11.

	Small	Medium	Large
Blue	17	47	63
Red	25	67	59

How many more red hoodies, in total, were ordered than blue hoodies?

5 Use the two-way table to work out how many female students chose French.

	Male	Female	Total
French	27		68
German	32	45	77

6 **R** Denis wants to know if James, Aran and India cycle, walk or drive to work.
Design a form to collect the information.

7 **R** A commuter wants to know whether the trains are late, on time or early for each day of the week.
Design a two-way table to record the data.

8 A mattress comes in three levels of comfort – soft, medium and firm.
The mattress can be covered in one of three fabrics – patterned, plain and textured.
The two-way table shows some information about the numbers of mattresses ordered in a shop one year.

	Soft	Medium	Firm	Total
Patterned		17	48	132
Plain		26	24	65
Textured	18		12	65
Total	100	78	84	

a How many soft patterned mattresses are ordered?

b How many soft plain mattresses are ordered?

c Copy and complete the two-way table.

9 **P** There are 20 people in a coffee shop.
Each person has tea or coffee.
8 of the people are male.
3 of the females had coffee.
A total of 12 people had tea.
Work out the number of males who had tea.

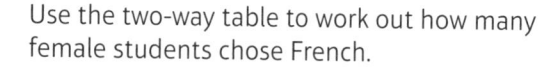
Example

Q9 hint Draw a two-way table.

10 **P** Anderley asked 40 people which TV channel they watched most the previous night – BBC, ITV or other.

Here is some information about her results.

12 of the 20 males watched BBC.

3 of the females watched ITV.

Of the 12 who said 'other', 5 were male.

Work out the number of women who watched BBC.

3.3 Representing data

1 Amy and Max record the number of houses they each sell over 3 months.

Example

	January	February	March
Amy	7	12	11
Max	4	15	9

a Draw a comparative bar chart for the data.

b In which month did they sell the most between them?

c Compare the house sales in January and February.

2

Exam-style question

The table shows the numbers of cats, dogs and rabbits seen by a vet over 2 days.

	Monday	Tuesday
Cats	5	2
Dogs	12	3
Rabbits	3	1

a Copy and complete the composite bar chart. **(4 marks)**

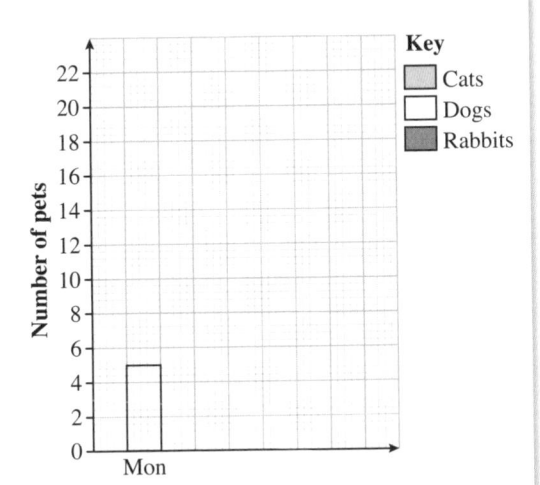

Key
☐ Cats
☐ Dogs
☐ Rabbits

b Compare the total number of appointments on Monday and Tuesday. **(2 marks)**

3 Ellie asked a group of people how many minutes they spend exercising each week. Her results are shown in the table.

Time, t (minutes)	Frequency
$0 \leqslant t < 30$	3
$30 \leqslant t < 60$	11
$60 \leqslant t < 90$	9
$90 \leqslant t < 120$	7

Draw a histogram to display this data.

4 **R** Here are a restaurant's takings over 3 days.

	Friday	Saturday	Sunday
Lunch	£350	£300	£450
Dinner	£400	£550	£150

Draw a chart for this data.

5 An online music shop recorded the numbers of CDs and downloads it sold over 5 years.

Year	1	2	3	4	5
CDs (1000s)	7	5.6	6.2	4.8	4.4
Downloads (1000s)	2.8	3.4	3.6	4	4.6

a Draw a set of axes. Put Year on the horizontal axis and Number sold on the vertical axis.

b Draw a line graph for the number of CDs sold. Use a different colour to draw a line graph for the number of downloads sold.

c Describe the trend in the number of CDs sold.
Describe the trend in the number of downloads sold.

6 Draw a line graph of the data in **Q4**.

> **Q6 hint** Draw two lines, one for lunch and one for dinner.

3.4 Time series

1 The graph shows the share price of a company after it has been floated on the Stock Market.

a What is the price of the shares after 4 days?

b Estimate how long the shares were worth less than 350 pence.

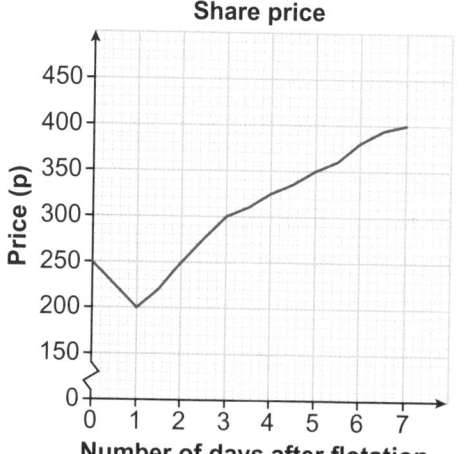

Share price

2 A baby's weight for 4 weeks is shown in the table.

Week	1	2	3	4
Weight (kg)	4.5	4.7	5.1	5.2

A nurse says, 'The baby's weight is increasing at a steady rate.'
Is he correct? Explain how you know.

3 The table shows the maximum daily temperature over a week in July.

Day	Mon	Tue	Wed	Thur	Fri	Sat	Sun
Temperature (°C)	19	21	22	25	22	19	21

Draw a time series graph to represent the data.

> **Q3 hint** Plot time along the horizontal axis and temperature on the vertical axis. The vertical axis needs to go up to 25. Join the points with straight lines.

4 A shop owner records the shop's monthly sales.
   January: £3400   February: £2100
   March: £2900   April: £2500
In May sales were twice those in January.
   a Construct a time series table for the shop.
   b Represent the data as a time series graph.

5 **R** A survey is carried out to find the numbers of people taking a holiday in the UK or abroad.

	2010	2011	2012	2013	2014
UK	35 000	49 000	45 000	42 000	40 000
Abroad	18 000	29 000	43 000	52 000	53 000

a Represent the data as two time series on a single graph.
b Describe the difference in the number of people taking holidays in the UK or abroad between 2010 and 2014.
c Abel predicts the number of people taking a holiday in the UK could decrease in 2015. Is he correct? Explain your answer.
d What is likely to happen to the number of people taking holidays abroad in 2015 and 2016?

3.5 Stem and leaf diagrams

1 The stem and leaf diagram shows the marks students scored in a test.

```
4 | 7  9  9
5 | 0  0  1  5  6  6
6 | 3  4  7  7
7 | 2
```

Key
4 | 7 represents 47%

a How many students scored less than 60%?
b What was the highest mark?

2 A travel agent recorded the ages (in years) of people booking an adventure holiday in one week.
The information is shown in the stem and leaf diagram.

```
1 | 8  8  9
2 | 0  0  0  3  4  5  7
3 | 0  0  2  5  7  9  9  9
4 | 0  1  1  1  3
```

Key
1 | 8 represents 18

a How many people booked the holiday?
b How many were younger than 30?
c What is the difference between the oldest and the youngest customers' ages?

3 Here are the times (in minutes) 15 students spent on their homework.

18, 19, 25, 33, 18, 19, 32, 39, 26, 27, 29, 30, 31, 34, 33

Construct a stem and leaf diagram to show this information.

Example

4

Here are the prices, in pounds (£), of 20 pairs of jeans.

35, 39, 41, 43, 44, 45, 22, 32, 29, 25, 46, 55, 59, 32, 22, 29, 28, 36, 25, 56

Show this information in an ordered stem and leaf diagram. **(3 marks)**

5 Here are the distances (in metres) some athletes ran in 20 seconds.

190, 192, 195, 184, 176, 171, 189, 193, 199, 170

Draw a stem and leaf diagram to display the data.

Q5 hint Use the 'hundreds' and 'tens' digits as the stem.

6 Angel recorded the times (in seconds) it took some members of a running club to run 100 m.

11.2, 12.2, 12.5, 13.7, 13.2, 12.9, 13.3, 14.4, 14.2, 12.9

Draw a stem and leaf diagram to display this data.

7 The weights of a group of patients at a hospital were recorded (in kg).

					Females		Males				
	9	9	4	3	3	6	0	1	1	2	7
7	5	2	1	1	0	7	5	7	7		
		7	2	0	0	8	1	3	9		
						9	2	2	5	8	

Key Females   Males

3 | 6   6 | 0

represents 63 kg   represents 60 kg

a What is the lowest male weight?

b What is the highest female weight?

c What is the difference between the lowest female weight and the highest male weight?

d Did more men than women weigh more than 80 kg? Explain your answer.

8 The manager of a fitness suite records the ages (in years) of people using the gym and the swimming pool.

Gym: 19, 23, 27, 34, 26, 29, 33, 39, 45, 51, 35, 25, 29, 39, 38, 42

Swimming pool: 43, 29, 39, 38, 52, 43, 39, 28, 17, 39, 61, 43, 45, 34, 44, 19

a Draw a back-to-back stem and leaf diagram to display the data.

b The manager says, 'The swimming pool attracts more younger people than the gym.' Is she correct? Explain your answer.

9 Charlie measures the heights of two sets of seedlings (in cm) kept in different parts of the greenhouse.

Position A	3.2	4.1	3.5	3.7	3.9	4.2	2.9	4.5	4.7
Position B	4.2	5.1	4.9	3.5	3.7	2.9	3.9	5.3	5.2

a Draw a back-to-back stem and leaf diagram to display the data.

b Did the seedlings grow better in Position A or Position B? Explain your answer.

3.6 Pie charts

1 A school surveyed its students for their opinion on how good their school lunches are. The pie charts show the results.

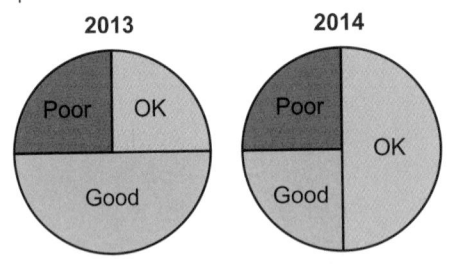

a What fraction of students thought the food was good in 2014?

b Compare the percentage of students who thought the food was OK in 2013 and 2014.

c Do you think the food has improved? Explain your answer.

2 A group of parents were asked whether they thought the local park's playground facilities were good. The results are shown in the pie chart.

Survey results

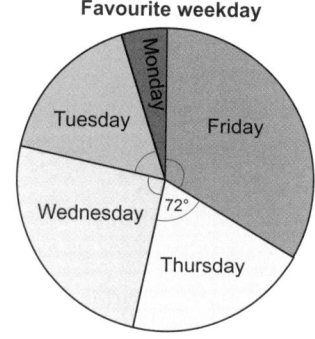

a What percentage of the parents said 'Don't know'?

b Measure the 'No' section with a protractor.

c Mr Ethan says, 'More than three times as many people said 'Yes' as 'No'.' Explain why he is correct.

3 **R** The pie charts show a student's punctuality over two different terms.

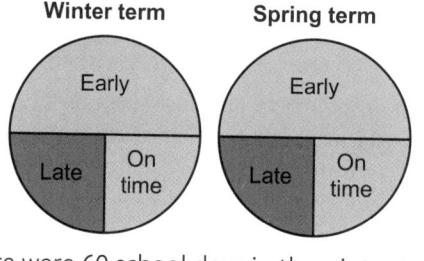

There were 60 school days in the winter term and 52 in the spring term.

a Calculate the number of days on which the student was late in the winter term.

b The form teacher says, 'This student was late the same number of days in each term.' Is she correct? Explain your answer.

4 A librarian records the items borrowed from the library one day.

The information is shown in the table.

Example

Item	Frequency
Classic fiction book	20
Contemporary fiction book	30
Non-fiction book	25
DVD	10
CD	5

a Work out the total frequency.

b Work out the angle for one item on a pie chart.

c Work out the angle for each type of item on a pie chart.

d Draw a pie chart to show the information.

5

Exam-style question

45 students choose to study a foreign language for GCSE.

The table shows the number of students who chose each language.

Language	Number of students
French	20
German	15
Spanish	9
Mandarin	1

Draw an accurate pie chart to show the information shown in the table. **(4 marks)**

6 20 people purchased a hoodie in a shop one day. They each chose one of four colours.

green, yellow, green, blue, red, red, red, red, blue, yellow, blue, green, blue, red, red, red, blue, green, red, yellow

Draw a pie chart to show this information.

7 **R** The pie chart shows the percentage of visits to the vet clinic by different animals in one week.

20 dogs visited the vet clinic.

Vet clinic visits

a How many animals visited the vet clinic in total?

b How many
   i cats ii rabbits visited the vet clinic?

8 A head teacher asked some students what their favourite weekday was. She drew a pie chart of her results.

Favourite weekday

a The head teacher knows that 20 students said 'Wednesday'. How many students were surveyed?

b What is the modal favourite weekday?

Q8a hint Measure the angle in the pie chart for Wednesday.

9 The owner of a car wash records the number of cars washed on three days.

Day	Monday	Tuesday	Wednesday
Number of cars	5	17	8

a Draw a bar chart for the data.

b Now draw a pie chart.

3.7 Scatter graphs

1 The age, in years, of some students and the time they took to run 100 m, in seconds, was recorded.

Age (years)	12	14	14	15	16	13
Time (s)	20	16	15	15	13	17

a Draw a scatter graph to display the data.

b Copy and complete the sentence.
As age increases, the time taken to run 100 m ....

2 Dr Swan recorded the ages of nine patients and the number of times they had visited him in the last year.

Age (years)	25	39	49	62	74	18	35	82	93
Number of visits	2	4	6	10	11	1	5	15	14

a Draw a scatter graph to display the data.

b Copy and complete the sentence.
As a person's age increases, the number of visits to the doctor ....

3 Look back at your graphs for **Q1** and **Q2**. Describe the type of correlation in each one.

4 Match each real-life example to the correct type of correlation.
a The relationship between height and weight
b The relationship between eye colour and weight
c The relationship between speed of train and time taken to travel 100 km

A No correlation
B Positive correlation
C Negative correlation

5

The lengths of time spent on a treadmill and the number of calories burned were recorded for 12 athletes. The data is shown in the scatter graph.

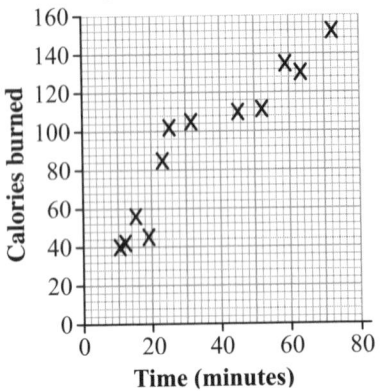

a Describe the relationship between the length of time spent on the treadmill and the number of calories burned. **(1 mark)**

b 4 squares of chocolate contain 120 calories. How many people did not burn off this number of calories? **(1 mark)**

Q5a hint This means write the type of correlation, so _____ correlation.

6 An ice cream vendor records the maximum temperature and the number of ice creams sold each day for a week. The table shows the results.

Temp. (°C)	25	20	22	27	29	31	22
No. of ice creams sold	100	73	90	140	152	170	149

a Plot the information on a scatter graph.

b One of the points is an outlier. Use your scatter graph to identify this outlier.

c Describe the correlation between temperature and the number of ice creams sold.

7 The table shows the ages of some students and the distance they live from school.

Age (years)	11	13	16	15	14	11	12	15	16
Distance (miles)	3.2	1.8	4.5	13.6	2.3	4.5	8.6	1.0	9.5

Plot the information on a scatter graph.

8 Copy and complete the table.
Tick the correct type of correlation for each set of data and whether you think one causes the other.

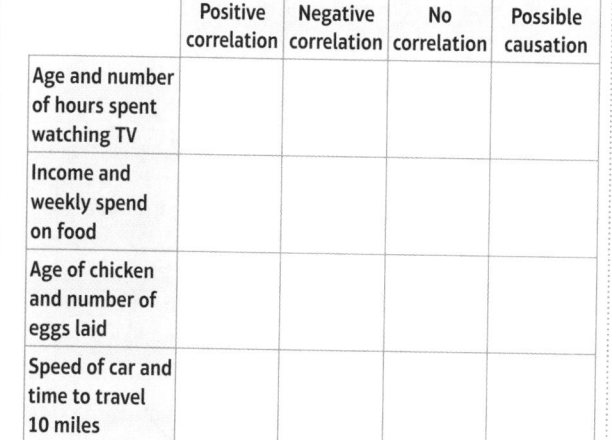

	Positive correlation	Negative correlation	No correlation	Possible causation
Age and number of hours spent watching TV				
Income and weekly spend on food				
Age of chicken and number of eggs laid				
Speed of car and time to travel 10 miles				

3.8 Line of best fit

1   A fisherman recorded the mass and length of six trout. The table shows his results.

Length (cm)	35	37	38	37	41	43
Mass (g)	600	755	810	720	870	930

   a   Draw a scatter graph for the data.
   b   Draw a line of best fit on your graph.

2   A school records the number of days some students are absent and the distance they live from school.

Number of days absent	3	8	1	4	7	3
Distance from school (miles)	18	2	9	3	7	4

   a   Draw a scatter graph for the data.
   b   Describe the correlation.
   c   Can you draw a line of best fit? Explain your answer.

3   **R** Some students took two maths papers, Calculator and Non-calculator. The table gives the percentage marks for seven students.

Example

Calculator (%)	56	73	49	90	47	76	54
Non-calculator (%)	55	69	45	91	78	69	55

   a   Plot the points on a scatter graph.
   b   Draw a line of best fit on the scatter graph.
   c   Are there any outliers? Explain your answer.
   d   Describe the relationship between the marks in the two papers.

   e   Eloise's mark in the Calculator paper is 65%. Estimate her mark in the Non-calculator paper.
   f   Phil's mark in the Non-calculator paper is 97%. Use your line of best fit to estimate his mark on the Calculator paper.

4   **Exam-style question**

   The scatter graph shows information about 10 apartments in a city.

   The graph shows the distance from the city centre and the monthly rent of each apartment.

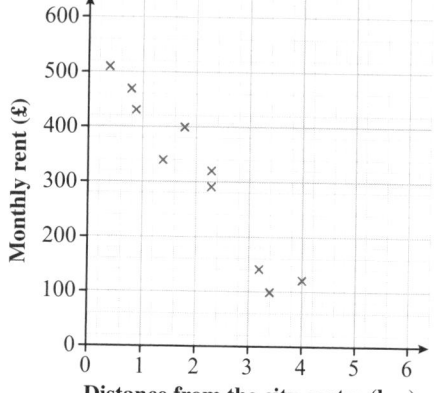

   a   Describe the relationship between the distance from the city centre and the monthly rent.   **(1 mark)**

   An apartment is 2.8 km from the city centre.

   b   Find an estimate for the monthly rent for this apartment.   **(2 marks)**

   March 2013, Q2, 1MA0/1H

3 Problem-solving

Solve problems using these strategies where appropriate:
•   **Use pictures**   •   **Use smaller numbers.**

1   **R** The chart shows the number of animals seen by the local vet this month. The total number of animals is 200.
   Draw a bar chart showing the same information.

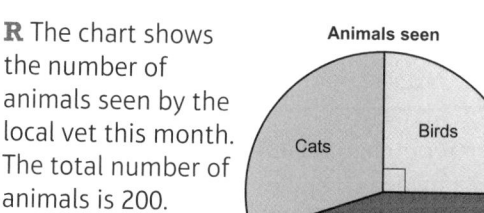

Animals seen

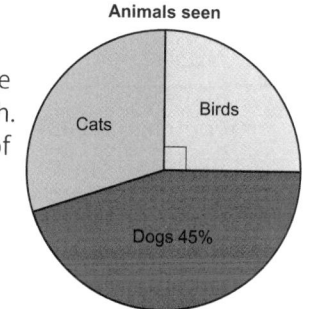

2 This is part of a bus timetable.

Service Number	310	316	310	314	310	316	310
Huddersfield Bus Station	1100	1115	1130	1145	1200	1215	1230
Honley Bridge New Mill Rd		1129		1159		1229	
Honley Bridge Woodhead Rd	1114		1144		1214		1244
Holme Valley Hospital	1120		1150		1220		1250
New Mill Holmfirth Rd				1206			
New Mill Greenhill Bank Rd		1136				1236	
Holmfirth Bus Station	1125	1147	1155	1214	1226	1248	1256
Scholes Moorlands Cres	1131		1201		1234		1304
Hepworth	1139		1209		1241		1311
Parkhead		1154				1255	
Holme				1229			

You arrive at Honley Bridge New Mill Road at 11.30 am.

a What is the earliest time you could get to Holmfirth Bus Station?

b How long will the bus ride take?

c What is the earliest time you could get to Parkhead?

You can walk from Honley Bridge New Mill Road to Honley Bridge Woodhead Road in 10 minutes.

d Could you get to either Holmfirth or Parkhead any earlier? Explain your answer.

3 A biscuit company is looking at the production of its three most popular biscuits. It records how many it makes each three months (rounded to the nearest 100 000).

	Jan–Mar	Apr–Jun	Jul–Sep	Oct–Dec
Plain Oaty Crunch	12 800 000	15 300 000	14 200 000	18 900 000
Chocolate Oaty Crunch	16 200 000	14 500 000	12 700 000	33 300 000
Caramel Oaty Crunch	13 400 000	11 600 000	19 100 000	25 400 000

Draw a graph to help you answer these questions.

a In which three months did the company make the most biscuits?

b In which three months did the company make the fewest biscuits?

c In which three-month period is the proportion of Plain Oaty Crunch the highest?

d Comment on how the number of biscuits made varies through the year.

> **Q3 hint** How could you write these millions as smaller numbers on your graph? How will this make it easier for you to read values from your graph?

4 **R** Temi has chosen to study maths, economics and philosophy for A levels.

There are 120 students in Year 12. The Venn diagram shows the students that have chosen at least one subject in common with Temi.

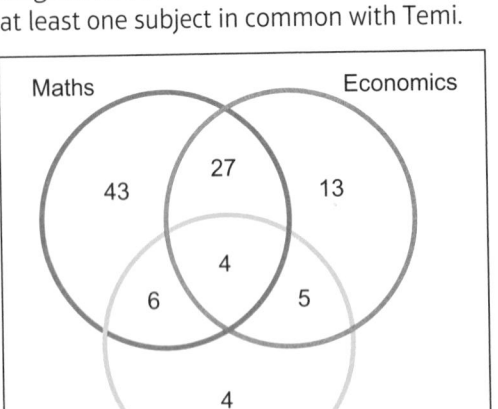
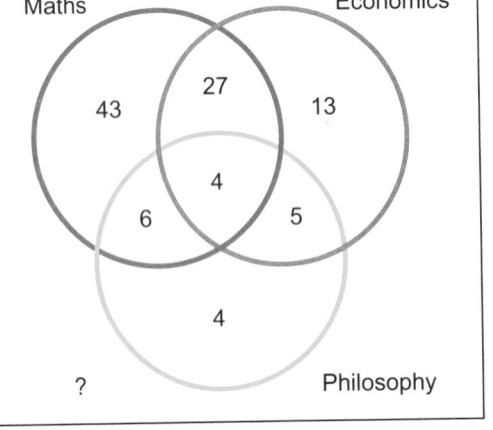

a How many students take exactly the same three subjects as Temi?

b What fraction of Year 12 take maths?

c How many students do not take any of the subjects that Temi takes?

d What percentage have only one subject in common with Temi?

e Why can you not represent this information in a two-way table?

5

Felicity asked 100 students how they came to school one day.

Each student walked or came by bicycle or came by car.

49 of the 100 students are girls.

10 of the girls came by car.

16 boys walked.

21 of the 41 students who came by bicycle are boys.

Work out the total number of students who walked to school. **(4 marks)**

November 2012, Q21, 1MA0/1F

6 **P** A vehicle hire company recorded the annual mileage of 25 vehicles.

21300 89614 21450 83199 30004
25600 94412 47708 67215 23118
48008 69401 13376 88583 93815
69030 96555 46917 40590 59780
72715 18168 42125 28288 42746

a Choose a suitable way to group this data.

b Represent your grouped data in a pie chart.

7 **R** a How does distance from school relate to journey time? Write two sentences to explain. Show your working.

Distance from school (km)	12	5	40	2	3	10	5	17	6	9	12	4	3	2	8
Journey time (minutes)	45	35	90	20	15	40	20	50	25	30	50	20	25	10	25

b The distances and journey times have been rounded. What have they each been rounded to?

4 FRACTIONS AND PERCENTAGES

4.1 Working with fractions

1 Write each pair of fractions with a common denominator.

a $\frac{1}{2}$ and $\frac{2}{3}$ b $\frac{7}{12}$ and $\frac{3}{4}$

2 **P** Millie has two bags of sweets, A and B.

For bag A, P(toffee) = $\frac{2}{3}$

For bag B, P(toffee) = $\frac{3}{5}$

From which bag is she more likely to choose a toffee sweet?

3 **R** Is $\frac{7}{12} > \frac{2}{3}$?

Show your working to explain your answer.

4

> **Exam-style question**
>
> Which of the two fractions $\frac{5}{6}$ and $\frac{3}{4}$ is the smaller?
>
> Explain how you know.
>
> You may use the grids to help with your explanation. **(3 marks)**

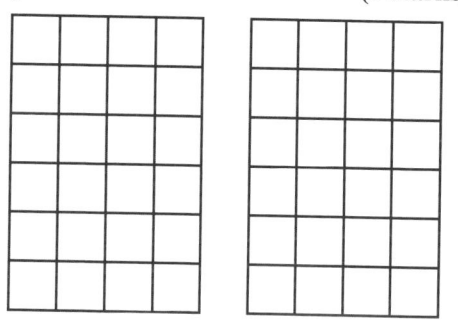

5 a Write in order of size, starting with the smallest.

$\frac{5}{8}, \frac{2}{3}, \frac{7}{12}$

b Write in order of size, starting with the largest.

$\frac{3}{4}, \frac{7}{10}, \frac{4}{5}$

Example

6 a **R** Predict which of these fractions is largest.

$\frac{5}{12}, \frac{7}{8}, \frac{7}{9}$

b Work out which fraction is largest. Was your prediction correct?

7 a **R** Write two fractions with the same numerator.

Explain how you can tell which one is larger.

b Explain which is smaller, $\frac{6}{7}$ or $\frac{7}{8}$

8 Work out these calculations. Give each answer in its simplest form.

Example

a $\frac{1}{2} + \frac{2}{5}$ b $\frac{2}{5} + \frac{3}{10}$

c $\frac{2}{3} - \frac{1}{2}$ d $\frac{3}{4} - \frac{1}{8}$

e $\frac{7}{8} - \frac{3}{16}$ f $\frac{1}{4} + \frac{3}{8}$

g $\frac{3}{5} - \frac{2}{15}$ h $\frac{11}{16} - \frac{1}{4}$

> **Q8 hint** Find the common denominator before adding or subtracting the fractions.

9 Work out

a $\frac{1}{4} + \frac{1}{5}$ b $\frac{1}{2} - \frac{1}{6}$ c $\frac{1}{3} + \frac{1}{8} - \frac{1}{9}$

10 **R** Rasheed is baking some cookies. It will take $\frac{1}{4}$ of an hour to make the mixture and $\frac{1}{3}$ of an hour to bake the cookies.

What fraction of an hour will this take in total?

11 R A group of students went to an outdoor centre.

$\frac{1}{4}$ of them went canoeing and $\frac{2}{5}$ of them went sailing. The rest went mountain biking.

What fraction of the group took part in water-based activities?

12 P The ancient Egyptians only used unit fractions.

For $\frac{5}{8}$ they wrote $\frac{1}{2} + \frac{1}{8}$

  a For which fraction did they write $\frac{1}{6} + \frac{1}{4}$?

  b How did they write $\frac{7}{10}$?

    Check your answer using a calculator.

13 Work out these. Give each answer in its simplest form.

  a $\frac{1}{12} + \frac{5}{8}$ b $\frac{1}{6} + \frac{7}{10}$ c $\frac{5}{6} - \frac{3}{8}$

  d $\frac{9}{10} - \frac{5}{6}$ e $\frac{1}{6} + \frac{3}{8}$ f $\frac{3}{4} - \frac{1}{6}$

  Check your answers using a calculator.

14 Work out

  a $1 - \frac{1}{5}$ b $1 - \frac{3}{8}$ c $2 - \frac{2}{3}$

4.2 Operations with fractions

1 Work out

  a $\frac{3}{5}$ of 20

  b $\frac{5}{6}$ of 30

  c $\frac{4}{7}$ of 42

Example

2 Find

  a $\frac{2}{3}$ of 300 m   b $\frac{5}{8}$ of 160 kg

3 A quiz has 60 questions.

Marius gets $\frac{4}{5}$ of the questions correct.

What does he score?

4 P Petra and Tom run a stall at a car boot sale. They split the money so that Petra gets $\frac{3}{5}$ and Tom gets the rest. They get £160 in total.

How much does Tom get?

5 Raj wants to put 6 filing cabinets along a wall in his office.

Each filing cabinet is 45 cm wide.

The wall is 2.4 m long.

Show that Raj cannot fit 6 filing cabinets along the wall.

6

> **Exam-style question**
>
> There are 12 000 people at a rock festival on a particular day.
>
> All of the people have a one-day pass, a two-day pass or are camping for all three days of the festival.
>
> $\frac{1}{4}$ of the people have a one-day pass.
>
> $\frac{3}{8}$ of the people have a two-day pass.
>
> Work out how many people are camping at the festival.   **(4 marks)**

> **Q6 hint** Check your answer gives a correct total number of people at the festival.

7 R The frequency table shows the numbers of people skating on an ice rink at a particular time.

People	Frequency
Men	25
Women	30
Boys	25
Girls	40

Lisa says, 'In a pie chart, the women will be represented by $\frac{1}{6}$ of its area.'

Explain why Lisa is wrong.

8 P 120 children were asked their favourite flavour of crisps.

This pie chart shows the results.

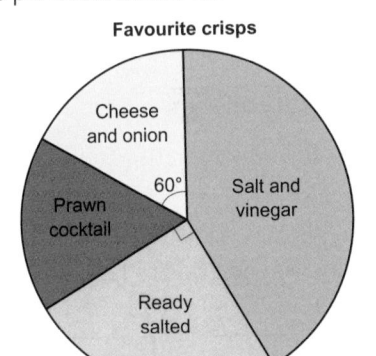

Favourite crisps

  a How many children chose cheese and onion?

  b Estimate the number of children who chose salt and vinegar.

9 Work these out. Give your answers as mixed numbers.

  a $\frac{5}{6} + \frac{3}{7} = \frac{\square}{42} + \frac{\square}{42} = \frac{\square}{42} = 1\frac{\square}{42}$

  b $\frac{7}{8} + \frac{3}{5}$ c $\frac{5}{6} + \frac{4}{5}$ d $\frac{5}{7} + \frac{5}{8}$

Check your answers using a calculator.

10 Work out

a $3\frac{1}{3} + 2\frac{1}{2}$ b $4\frac{3}{4} + 3\frac{1}{2}$

c $7\frac{1}{4} + 6\frac{1}{2}$ d $9\frac{1}{2} + 4\frac{2}{3}$

e $2\frac{1}{2} + 4\frac{1}{6}$ f $5\frac{3}{5} + 4\frac{1}{2}$

Example

11 Work out

a $4\frac{3}{4} - 3\frac{1}{3}$ b $8\frac{1}{2} - 4\frac{2}{3}$ c $12\frac{1}{2} - 9\frac{3}{4}$

d $8 - 2\frac{1}{3} - 3\frac{3}{5}$ e $7\frac{1}{2} - 2\frac{3}{4}$ f $9\frac{3}{5} - 6\frac{1}{2}$

12 **P** In this diagram, the number in each box is the sum of the two numbers below it.
Find the missing numbers.

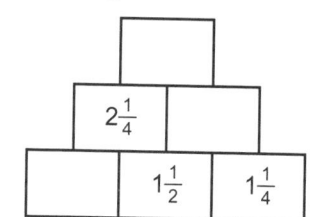

4.3 Multiplying fractions

1 Copy and complete.

a $8 \times \frac{1}{3} = \frac{8 \times 1}{3} = \frac{\square}{3} = \square\frac{\square}{3}$

b $8 \times \frac{5}{3} = \frac{\square}{3} = \square\frac{\square}{3}$

2 Work these out. Write your answers as mixed numbers if necessary.

a $3 \times \frac{7}{3}$ b $4 \times \frac{5}{2}$ c $6 \times \frac{7}{5}$

d $\frac{7}{4} \times 7$ e $\frac{12}{5} \times 3$

3 Sally has got 5 trifles to make. Each trifle will use $\frac{2}{3}$ pint of cream.
Show that Sally will need $3\frac{1}{3}$ pints of cream altogether.

4 **R** Yessim is selling his old laptop for £240.
He reduces the price by $\frac{1}{5}$.
What is the new price?

5 Substitute $x = 5$ and $y = 3$ into

a $\frac{1}{2}(x + 5)$ b $\frac{1}{3}(y + 6)$

c $\frac{1}{4}(x + 7)$ d $\frac{1}{2}(x - y)$

6 A mortar mix uses $\frac{3}{5}$ kg of sand for each kilogram of mortar.
How much sand is needed for 10 kg of mortar?

7 Work out

a $\frac{1}{3} \times \frac{1}{4}$ b $\frac{1}{2} \times \frac{1}{5}$

c $\frac{2}{5} \times \frac{1}{4}$ d $\frac{1}{6} \times \frac{3}{5}$

e $\frac{5}{6} \times \frac{2}{3}$ f $\frac{3}{7} \times \frac{4}{5}$

g $\frac{4}{9} \times \frac{3}{8}$ h $\frac{3}{4} \times \frac{8}{9}$

Example

8 **P** A footballer donates $\frac{1}{5}$ of his income to charity.
He makes £$\frac{3}{4}$ million in one particular season.
How much money does he donate to charity that season?

9 Work out

a $\frac{3}{5} \times \frac{3}{4}$ b $\frac{7}{12} \times \frac{4}{5}$

c $\frac{5}{7} \times \frac{8}{15}$ d $\frac{16}{21} \times \frac{7}{8}$

e $\frac{11}{16} \times \frac{8}{15}$ f $\frac{7}{16} \times \frac{4}{21}$

Example

10 In a running club, $\frac{2}{3}$ of the members are male.
Of these, $\frac{1}{4}$ are under 16.
What fraction of the running club are male and under 16?

11 Dylan buys $\frac{5}{8}$ tonne of coal. He gives his mother $\frac{2}{5}$ of it.
Dylan needs 350 kg of coal to last for a month. Does he have enough?
Show how you worked it out.

12 **Exam-style question**

A concrete mixer can mix a maximum of 90 litres at a time.

If the mixer is two-thirds full, work out how much concrete (in litres) is being mixed. **(2 marks)**

13 Work out

a $4 \times 2\frac{2}{3}$ b $5 \times 1\frac{3}{7}$

c $3\frac{3}{5} \times 8$ d $4\frac{2}{3} \times 12$

Q13a hint Write $2\frac{2}{3}$ as an improper fraction first, then multiply by 4.

14 Jamie trains for $1\frac{1}{4}$ hours every day for a marathon.
How long does he spend training each week (Monday to Friday)?

15 Briony needs 8 pieces of curtain material $2\frac{1}{4}$ m in length. She has a 19 m length of material.
Does she have enough?

16 **P** Amy's parents need a mortgage to buy a house. They are offered two options.

Option 1: Two and a half times their joint salary

Option 2: Three and a half times the larger salary and two times the smaller salary.

Their salaries are £36 000 and £43 000.

Show which option would give them the larger mortgage.

4.4 Dividing fractions

1 Write down the reciprocal of

a $\frac{2}{3}$ b $\frac{1}{7}$ c 6

2 Work out

Example

a $5 \div \frac{1}{4} = 5 \times \square = \square$

b $6 \div \frac{1}{3}$ c $8 \div \frac{1}{6}$

d $5 \div \frac{3}{5}$

3 How many $\frac{2}{3}$-litre soup bowls can you fill from a 3-litre saucepan of soup?

4 Work out these divisions. Write your answer as a mixed number if necessary.

a $7 \div \frac{5}{2}$ b $10 \div \frac{5}{4}$ c $12 \div \frac{15}{4}$

5 Work out these divisions. Write your answer as a mixed number if necessary.

a $9 \div 2\frac{1}{3} = 9 \div \frac{\square}{3} =$

b $12 \div 3\frac{1}{5}$ c $8 \div 4\frac{2}{3}$

6 Work out

a $\frac{1}{4} \div 3$ b $\frac{3}{5} \div 4$

c $\frac{7}{4} \div 8$ d $2\frac{4}{5} \div 7$

7 Dave cycles $3\frac{3}{4}$ km to work.

How far has he cycled when he is halfway? Give your answer in metres.

Check your answer using an inverse operation.

8 Leanne takes $3\frac{3}{4}$ hours to do 5 tasks.

Each task takes the same time.

What fraction of an hour does 1 task take?

9 Exam-style question

Zara has 15 m of ribbon. She needs $\frac{5}{6}$ m of ribbon to make a wedding decoration.

How many wedding decorations can she make from this ribbon? **(2 marks)**

Q9 hint Check your answer using the inverse operation.

10 Work out these divisions. Write your answer as a mixed number if necessary.

a $\frac{1}{4} \div \frac{1}{3} = \frac{1}{4} \times 3 = \dfrac{\square}{\square}$

b $\frac{5}{6} \div \frac{1}{3}$ c $\frac{7}{12} \div \frac{1}{4}$

4.5 Fractions and decimals

1 Exam-style question

Example

Write $\frac{9}{20}$ as a decimal. **(1 mark)**

2 Copy and complete this table.

Decimal	0.001	0.01	0.1	0.125	0.2	0.25	0.5
Fraction		$\frac{1}{100}$					$\frac{1}{2}$

3 Write these fractions as decimals. Use your table from **Q2** to help you.

a $\frac{3}{5} = 3 \times \frac{1}{5} = 3 \times \square =$

b $\frac{7}{8}$ c $\frac{9}{2}$ d $\frac{11}{4}$

4 Work out

a $\frac{1}{5}$ of 20 = 0.2 × 20 =

b $\frac{3}{4}$ of 24   c $\frac{3}{2}$ of 12   d $\frac{7}{10}$ of 400 m

Q4 hint Converting a fraction to a decimal can make a calculation easier.

5 Which of these fractions is closest to $\frac{1}{2}$?

$\frac{9}{20}, \frac{7}{12}, \frac{11}{15}, \frac{5}{8}$

6 Write these decimals as fractions in their simplest form.

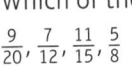

Example

a 0.77   b 0.56

c 0.433   d 0.650

e 0.072

7 Write these in order of size, starting with the smallest.

a $\frac{7}{9}$, 0.72, $\frac{3}{4}$, 0.7

b $\frac{1}{3}$, 0.3, $\frac{3}{8}$, 0.34

c $\frac{7}{2}$, −3.6, $-\frac{15}{4}$, 2.9

8 **R** Jamilla says that $\frac{1}{9}$ of 350 m is 38.89 m rounded to the nearest centimetre.

Show that Jamilla is correct.

9 **R** Jill and Claire share an £80 restaurant bill.
Jill pays $\frac{5}{6}$ and Claire pays $\frac{1}{6}$.
a How much does each pay?
b Will the amounts in your answers to part **a** pay the whole bill?
Should you round up or down when sharing a bill?

10 **R** Ruby uses this formula to work out the time needed to cook a joint of meat.

Time (hours) = $\frac{10 \times \text{mass (kg)}}{9}$

A joint of meat weighs 6 kg.
Work out the cooking time. Give your answer in hours and minutes.

11 Alex draws a pie chart to show the ways students travel to school.

Travel to school

Show that the fraction of students who come by bus is $\frac{12}{36}$.

12 Fiona's bus fare to work was £1.50. It goes up to £1.80.
Write the new bus fare as a fraction of the old bus fare.
Give your answer in its simplest form.

13 Luke buys a house for £150 000. He sells it several years later for £200 000.
Write the selling price as a fraction of the original price in its simplest form.

14 Write 24 minutes as a fraction of an hour in its simplest form.

4.6 Fractions and percentages

1 Write as a fraction in its simplest form
a 12%   b 35%   c 55%
d 84%   e 4%

2 95% of students have a mobile phone.
Write this percentage as a fraction in its simplest form.

3 Show that 18% is smaller than $\frac{1}{5}$.

4 A group of 20 students chose a sporting activity.
8 chose tennis. The rest chose volleyball.
What percentage chose
a tennis
b volleyball?
c Add together your answers to parts **a** and **b**. Write a calculation to work out the percentage who chose volleyball, using the percentage that chose tennis.

Example

5 **R** A company wants to give a reduction of between 30% and 40% on its bicycles.
Which of these fractions could they use?
$\frac{9}{20}, \frac{1}{4}, \frac{3}{8}$

Show your working to explain your answer.

6 **R** Danny says that $\frac{1}{6}$ is 15%.
Show that Danny is wrong.

7 Hasan does a survey on people's favourite type of music.
He draws a bar chart to show his results.

Favourite types of music

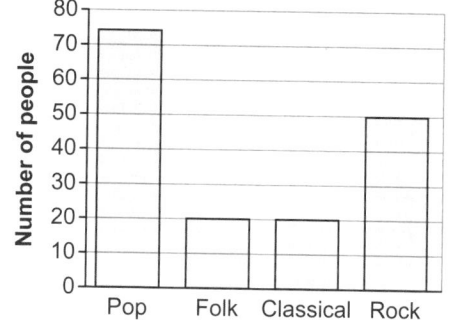

a What percentage of people chose rock music?
Hasan says that half as many people like folk and classical music as like pop.
b Show that Hasan is incorrect.

8 Write
  a 35 as a percentage of 50
  b 520 as a percentage of 800
  c 75p as a percentage of £2.50
  d 900 m as a percentage of 3.6 km

  > Q8c, d hint Write both amounts in the same units.

9 **R** Elsa got 34 out of 40 in a quiz. Marcus scored 80% in the same quiz.
  Who achieved the higher score? Explain your answer.

10 **R** Harvey took a chemistry test and a biology test.
  He scored 56 out of 70 for chemistry and 65 out of 80 for biology.
  In which test did Harvey score the lower percentage?

11 There are 270 students in Year 11 at a school. 175 of them want to study A levels.
  What percentage of the students want to study A levels?
  Round your answer to the nearest whole number.

12 **Exam-style question**

  A TV channel shows 225 programmes in the course of a week. 21 of these are news programmes.

  What percentage of the programmes are news programmes?

  Give your answer to 1 decimal place.

  (3 marks)

4.7 Calculating percentages 1

1 Write these percentages as decimals.
  a 35%   b 28%

2 The diagram shows a fraction–decimal–percentage triangle.

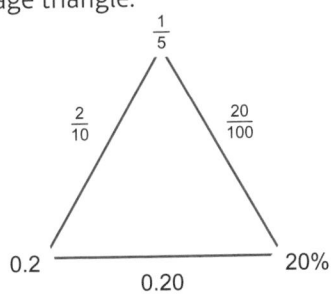

Draw fraction–decimal–percentage triangles for
  a $\frac{3}{4}$ b 40%
  c 0.7   d $\frac{5}{8}$

3 Write in order of size, starting with the largest.
  a $\frac{3}{4}$, 0.7, $\frac{13}{20}$, 78%   b $\frac{1}{3}$, 3.5%, 0.3, $\frac{2}{5}$

  > Q3 hint Convert to decimals.

4 A credit card has an interest rate of 14.9%. Write this percentage as a decimal.

5 a Write 25% as a fraction.
  b Work out 25% of 2500.

6 Work out
  a 80% of 600   b 30% of 700
  c 40% of 800

7 a Write 40% as a decimal.
  b Work out 40% of 500.

8 Work out these percentages by converting to a decimal first.
  a 70% of 350   b 65% of 400
  c 15% of 260

9 Judith buys a dress originally costing £80 reduced by 45% in a sale.
  How much money does Judith save?

10 James wants to buy some new trainers costing £65.99.
  He has two vouchers. Which voucher should James use? Explain your answer.

  £10 off if you spend £50 or more!

  20% off selected trainers

11 On Saturday a shoe shop sells 90 pairs of children's shoes.
  30% of the children's shoes sold are boys' shoes. How many pairs of children's shoes are not boys' shoes?

12 **P** Alima's family own a shop. They have to pay an annual business council tax of £1265.60. A 10% discount is applied as they are a new business.
  The council says they should pay £94.92 each month for the year.
  Check that the council are charging correctly. Show your working.

13 **P** Leanne sells a dining table at an auction.
She pays £7.50 to put the table in the auction and 8% of the selling price as a fee.
The table sells for £175.
How much does Leanne have to pay?

14 Write as a decimal
  a 160%   b 115%   c 430%

15 Petrol costs $2\frac{1}{2}$ times as much today as it did 20 years ago.
What percentage has it increased by?

16 Work out
  a 120% of £60   b 175% of 300 m
  c 350% of 400 kg

17 An antiques dealer buys a grandfather clock for £6000. He sells it at 140% of this cost.
How much did the antiques dealer sell the clock for?

18 A greengrocer buys onions for 36p per kg and sells them at 275% of the cost price.
How much does the greengrocer sell 1200 g of onions for? Write your answer to the nearest penny.

19 Find the simple interest when
  a £4000 is invested at 3.75% per annum (p.a.) for one year
  b £750 is invested at 2.25% p.a. for 4 years
  c £6000 is invested at 3.2% p.a. for 30 months.

Example

20 Ella's grandfather gives her £2500. She saves the money in a bank account with a simple interest rate of 3.4% per annum.
How much money will Ella have in the bank account after $3\frac{1}{2}$ years?

21 **P** The cost of living increased by 30% from 2004 to 2014.
In 2004, the Smith family's weekly income was £420.
What would you expect their weekly income to be in 2014 if it increased at the same rate as the cost of living?

22 Rashid buys a car for £15 900.
The value of the car depreciates by 14% each year.
Work out the value of the car at the end of the first year he has owned it.

Q22 hint 'Depreciates' means that the value of the car decreases.

23
Exam-style question

Alice earns an income of £1340 each month.
She spends 45% of her income on rent.
She spends $\frac{1}{3}$ of her income on food and bills.
Work out how much she has left each month. **(2 marks)**

4.8 Calculating percentages 2

1 Increase
  a 45 by 40%
  b 300 by 67%
  c 4000 by 0.3%

Example

2 Danny works part time for £54 per week. His employer agrees to raise his weekly wage by 4%.
What is Danny's new weekly wage?

3 Decrease
  a 65 by 20%
  b 800 by 35%
  c 3000 by 0.7%

Example

4 **P** A sofa company reduces its prices by 35%.
A sofa was £600.
What is the new price of the sofa?

5 **P** Mia wants to improve her 1500 m running time by 5%.
Her current time is 6 minutes.
What is her new target time?

6 The cost of sending a parcel is increasing by 20%. A 2 kg parcel costs £12 to send.
  a Work out 20% of £12.
  b Work out the new price to send the parcel.
  c Work out 120% of £12.
  d What do you notice about your answers to parts **b** and **c**? Explain.

7 Write the decimal multiplier you can use to work out an increase of
  a 30%   b 64%   c 2.5%

8 Increase
  a 160 by 150%
  b 600 by 300%
  c 4000 by 7.5%

9 Write the decimal multiplier you can use to work out a decrease of
a 30%
b 45%
c 4%
d 7.5%

10 Decrease
a 400 by 30%
b 700 by 65%
c 160 by 15%
d 4000 by 7.5%

11 **P** A 5% discount is deducted from the cost of a meal as a special offer.
A meal cost £65. What is the total charge?

12 **Exam-style question**

Gemma is buying shampoo for her hair salon.
She finds out the cost at two suppliers.

Best Buy Mart

Shampoo offer – bulk buy!
Normal price £1.50 a bottle
Buy 12 bottles for the price of 9

Super Shop

Shampoo Special Offer
Normal price 96p
15% off the normal price

Gemma needs 60 bottles of shampoo.
She wants to buy it all from the same supplier, at the cheapest possible total price.
Which of the two suppliers should Gemma buy the shampoo from?   **(6 marks)**

Exam hint
Make sure you make a statement at the end that includes the cost of the shampoo and which shop Gemma should buy it from.

13 **R** Amit is self-employed. Last year, he earned £28 640.
He does not pay income tax on the first £10 000 he earned.
He pays tax of 20% for each pound he earned above £10 000.
How much tax must he pay?

14 **P** Sergei gets two quotes for roof repairs.
Quote 1: £780 including VAT
Quote 2: £675 excluding VAT
Which is the more expensive quote? Show your working.

Q14 hint VAT is charged at a rate of 20%. It is added to the bill in addition to the cost of the work.

15 **P** Aisha buys 3 tickets for a concert.
The cost of one ticket is £68 plus VAT.
VAT is 20%. Work out the total cost of the tickets.

16 **P** Alex is planning a raffle.
Each winner will get a gift voucher costing £30 plus VAT at 20%.
Alex has up to £200 to spend on prizes.
How many gift voucher prizes will there be?

17 **P** A family of 2 adults and 3 children are going on holiday.
The adult price for the holiday is £480.
The company is offering a discount of $\frac{1}{4}$ off adult prices, and a child costs 35% of the original adult price.
Work out the total cost of the holiday.

18 **R** A sports centre manager predicts that there will be a 30% increase in membership during January.
There are currently 650 members of the leisure centre.
How many members does the manager predict there will be by the end of January?

4 Problem-solving

Solve problems using these strategies where appropriate:
• **Use pictures**
• **Use smaller numbers**
• **Use bar models.**

Example

1 **R** Show that $\frac{3}{5}$ < 70%.

Q1 hint Use a bar model. Divide the bar into 10% sections.

2 Sarah, Abz and Nadine share a flat. The rent is £720 a month. Sarah gets the largest room. She pays $\frac{3}{8}$ of the rent. Abz pays 60% of the rest of the rent and Nadine 40%.
Work out how much Abz and Nadine each pay.

3 **R** Michael scores 19 out of 25 in a science test, 78% in a history test and only gets 4 wrong in a maths test out of 20.
Which subject is he performing best in? Explain your answer.

4 **R** Adam is 4 years older than Sam. Charles is twice as old as Sam.
Use s to represent Sam's age.
  a Write an expression for the sum of their ages.
The sum of their ages is 36 years.
  b Write an equation for the sum of their ages.
  c How old are Sam, Adam and Charles?

5 **R** Marta earns £28 225 a year. She is offered four choices for a pay rise.
  A No increase in year 1, then 15% in year 2
  B 8% in year 1 with no increase in year 2
  C 5% in year 1, then another 5% in year 2
  D 6% in year 1, then 4% in year 2
Which should she choose? Explain your answer.

> **Q5 hint** How can you use simple numbers to help you?

6 **R** Henry says, 'I have 25 g of salt in 2000 g of this salt solution.'
  a What is the percentage of salt in his salt solution?
He wanted the solution to have a 0.5% concentration.
  b Does he have too much or too little salt in the solution? Explain.
  c How much salt should have been in the solution?

7 **Exam-style question**

Liam borrows £12 350 to buy a car. He pays the money back with simple interest at a rate of 14% over the next 5 years.
  a How much interest will he pay on the loan? **(3 marks)**
  b How much would he save if he paid off the loan in 3 years instead of 5? **(2 marks)**

8 In a class of 25 students, 5 students don't have a dog or a cat.
Of the rest, 13 have a dog and 11 have a cat.
How many have only a cat?

9 **R** Paul and Avish are mixing drinks for their friends. They pour $1\frac{1}{2}$ cups of orange squash and $\frac{3}{5}$ of a cup of pineapple squash into a 6-cup jug. They then fill the jug up with water.
  a How much water is in the jug?
  b Avish says the amount of water is nearly double the amount of squash. Is he right? Explain your answer.

10 Loré gives Rose and Margaret all her bottles of nail varnish.
Rose gets $\frac{7}{20}$ of them. This is 24 fewer than Margaret gets.
  a How many bottles of nail varnish were there to begin with?
Margaret sells $\frac{4}{13}$ of hers for £2.50 a bottle.
  b What percentage of Loré's set did Margaret sell?
  c How much did she make?

11 This is a plan of Katie's bedroom.

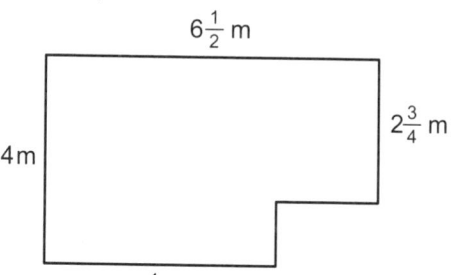

She wants to buy coving to line all around the edges of the ceiling.
  a What length of coving does she need?
She also wants to tile the floor. The tiles are 50 cm by 50 cm.
  b How many tiles will she need to cover her floor?

5 EQUATIONS, INEQUALITIES AND SEQUENCES

5.1 Solving equations 1

1 Copy and complete to solve this equation.

$$\frac{x}{5} = 9$$

$x = 9 \times \square$

$x = \square$

2 Solve

a $\frac{t}{3} = 4$　　b $\frac{m}{5} = 7$　　c $\frac{r}{3} = 6$

d $\frac{t}{5} = 5$　　e $\frac{m}{4} = 3$　　f $\frac{b}{2} = 12$

3 Copy and complete to solve these equations using a balancing method.

Example

a
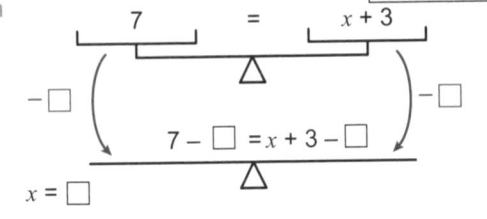

$x = \square$

b
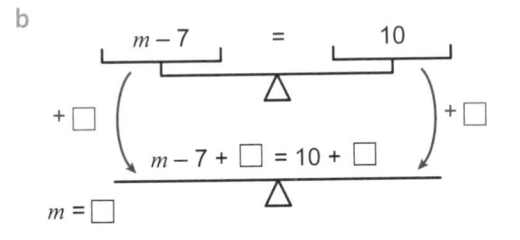

$m = \square$

c
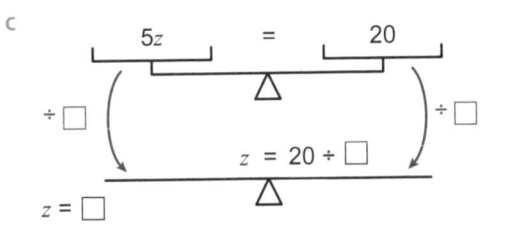

$z = \square$

d
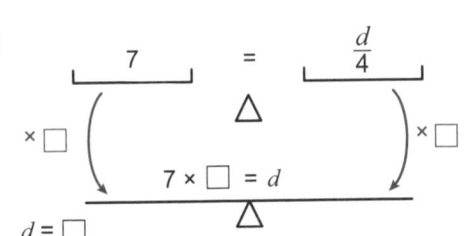

$d = \square$

4 Use a balancing method to solve

a $m + 8 = 12$　　b $y - 5 = 14$

c $x + 11 = 12$　　d $5 + r = 20$

5 Simplify the left-hand side, then solve.

a $4x + 6x = 30$　　b $9y - 4y = 15$

c $8m - 7m = 12$　　d $3q + q = 24$

6 **R** a Write an equation for the sum of these angles.

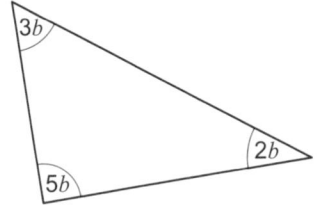

b Solve your equation to find the value of b.

7 **R** The perimeter of this isosceles triangle is 32 cm.

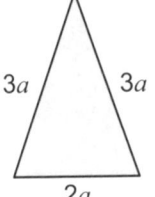

a Write an equation for the perimeter.

b Work out the value of a.

c Work out the length of the shortest side.

8 **R** Calculate the size of the smallest angle.

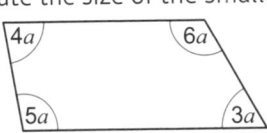

> **Q8 hint** What do the angles in a quadrilateral add up to?

9 **Exam-style question**

The perimeter of this pentagon is 56 cm.

Work out the length of the sides marked x and $2x$.

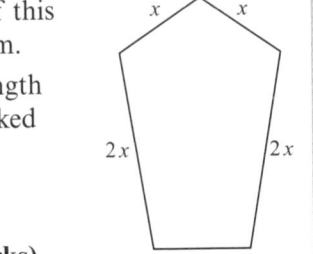

(4 marks)

10 Amy works x hours and earns £12 per hour. One day she is paid £84.

   a Write an equation involving x.

   b Solve your equation to work out how many hours Amy worked.

11 Yogini eats 12 chocolates from a box containing y chocolates. There are now 18 chocolates in the box.

   a Write an equation involving y.

   b Solve your equation to find the original number of chocolates in the box.

5.2 Solving equations 2

1 Copy and complete these inverse function machines.

   a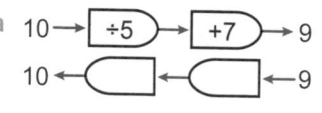

   b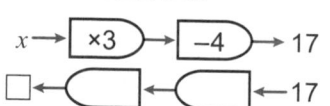

2 a Solve the equation $3x - 4 = 17$ using this function machine.

   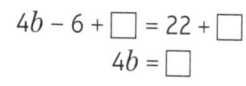

   b Solve these equations using function machines.

   i $4x + 5 = 13$

   ii $2m - 7 = 11$

   iii $8m + 10 = 30$

3 Copy and complete.

   Example

   a $\quad 4b - 6 = 22$
   $4b - 6 + \square = 22 + \square$
   $\quad\quad 4b = \square$
   $\quad\quad\; b = \square$

   b $\quad 9m + 3 = 21$
   $9m + 3 - \square = 21 - \square$
   $\quad\quad 9m = \square$
   $\quad\quad\; m = \square$

4 Solve these equations.

   a $3x + 4 = 10$ b $5x - 7 = 8$
   c $2x - 6 = 14$ d $6x + 4 = 10$
   e $2x + 2 = 3$ f $7x + 3 = 9$
   g $2a + 3 = -3$ h $-2x + 4 = -2$
   i $a + 7 = -1$

R = Reasoning   P = Problem-solving   5.2 Solving equations 2

5

   a Solve $m + 4 = 12$ **(1 mark)**

   b Solve $\dfrac{p}{5} = 7$ **(1 mark)**

   c Solve $3x + 4 = 12$ **(2 marks)**

6 **R** I think of a number. I divide it by 3 and add 7. The result is 12. Find the number.

7 **R** a Write an equation for the diagram.

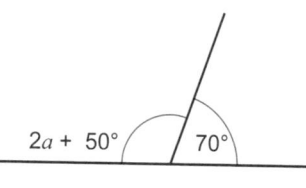

   b Solve your equation to find the value of a.

8 **R** The sizes of the angles in a triangle are $2a°$, $6a - 30°$ and $10° + 2a°$.

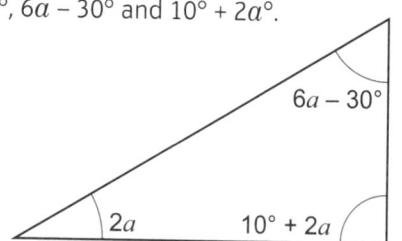

Find the value of a.

9 **R** The length of each side of a regular pentagon is $(a + 5)$ cm. The perimeter of the pentagon is 40 cm. What is the value of a?

10 **R** India has 4 more CDs than Jarod. Together they have 18 CDs. How many CDs does India have?

11 Solve these equations.

   a $\dfrac{2x}{3} = 10$ b $\dfrac{3b}{7} = 9$

   c $\dfrac{4m}{5} = -10$ d $-\dfrac{2x}{7} = 4$

   Q11a hint

   $x \rightarrow \boxed{\times 2} \rightarrow \boxed{\div 3} \rightarrow 10$

12 Solve these equations.

   a $\dfrac{a}{3} + 1 = 4$ b $\dfrac{m}{2} - 5 = 7$

   c $\dfrac{x}{3} - 2 = 12$ d $\dfrac{d}{6} + 5 = 11$

   e $\dfrac{k}{5} + 3 = 7$ f $\dfrac{b}{4} - 5 = -2$

13 Solve

a $\frac{1}{2}t = 10$ b $\frac{p}{12} = 10$

c $\frac{3}{4}m = 18$ d $\frac{2}{3}x = 10$

e $\frac{4x}{5} = 12$ f $\frac{b}{3} = \frac{1}{4}$

5.3 Solving equations with brackets

1 Expand and solve

a $2(b + 4) = 12$ b $6(m + 2) = 24$

c $5(x + 1) = 30$ d $2(x - 3) = 20$

e $9(b - 10) = 18$ f $5(m - 2) = 100$

g $10(3 - f) = 20$ h $-3(x + 1) = -9$

i $-4(10 - x) = -36$

2 Solve

a $\frac{2x + 1}{3} = 3$ b $\frac{3b - 4}{7} = 2$

c $\frac{2x + 5}{9} = 3$ d $\frac{2m - 6}{2} = 7$

3 Solve these equations.

a $10x = 8x + 12$

b $5x = 2x + 36$

c $2c + 7 = 3c + 2$

d $4b - 30 = b + 3$

Example

4 Solve these equations.

a $3a + 10 = 7a - 2$ b $2a + 5 = a + 3$

c $2b + 7 = 5b - 23$ d $3m - 10 = m + 1$

e $5x - 2 = 2x - 4$ f $3h + 5 = 5h - 2$

5 Solve these equations.

a $20 - 3x = 2$ b $10 - 2x = 5x + 3$

c $3 - 2x = 9 + x$ d $10 + 2x = 20 - 3x$

e $5x - 4 = 10 - 2x$ f $8x + 1 = 5 - x$

6 Exam-style question

Solve $4(x - 1) = 2x + 14$ **(4 marks)**

7 **R** Lawrence is paid 5 times as much as Ethan.
He also earns £40 more than Ethan.
How much do they each earn?

Q7 hint Write an equation and solve it.
Use x for Ethan's earnings.

8 Expand the brackets on both sides, then solve.

a $3(b + 7) = 12(b + 1)$

b $2(m - 3) = 4(6 - m)$

c $4(3 + m) = 6(m - 2)$

d $4(x + 1) = 3(3 + x)$

e $10(10 + x) = 12(2x + 6)$

f $5(2m + 7) = 5(5m - 2)$

9 a Work out the lengths of the sides of the square.

$(20 - 4x)\,\text{cm}$

$(6x - 10)\,\text{cm}$

b What is the perimeter?

10 **P** A rectangle has width 5 cm, and the length is y cm more.
The area of the rectangle is $40\,\text{cm}^2$.

$(y + 5)\,\text{cm}$

5 cm

a Write an equation and solve it to find y.

b What is the perimeter of the rectangle?

5.4 Introducing inequalities

1 Draw six number lines from −5 to +5. Show these inequalities.

a $x < 4$ b $x > -2$

c $x < -0.5$ d $x > -3$

e $x \leqslant 3\frac{1}{2}$ f $x \geqslant -4$

Example

2 Draw six number lines from −5 to +5. Show these inequalities.

a $-2 < x < 3$ b $-3 < x \leqslant 0$

c $1 \leqslant x \leqslant 4$ d $-4 < x \leqslant -1$

e $-\frac{1}{2} < x \leqslant 5$ f $-3 \leqslant x < -2$

3 Write down the inequalities represented on these number lines.

a
```
   -5 -4 -3 -2 -1  0  1  2  3  4  5
```
→ x

b
x
```
   -5 -4 -3 -2 -1  0  1  2  3  4  5
```
→ x

c
```
   -5 -4 -3 -2 -1  0  1  2  3  4  5
```
→ x

d
```
   -5 -4 -3 -2 -1  0  1  2  3  4  5
```
→ x

e
```
   -5 -4 -3 -2 -1  0  1  2  3  4  5
```
→ x

f
```
   -5 -4 -3 -2 -1  0  1  2  3  4  5
```
→ x

4 When exercising, Mr Williams's heart rate, h, is between 185 and 190 bpm (beats per minute) inclusive.
Show this on a number line.

5 Write down the integer values of x that satisfy each of these inequalities.
a $3 < x < 10$ b $-2 < x \leqslant -1$
c $-2 \leqslant x \leqslant 2$ d $1.5 < x < 6$
e $-2\frac{1}{2} < x \leqslant 4$ f $-1 < x < 1$

6 $n \geqslant 3$
Write an inequality for
a $3n$ b $n + 1$
c $n - 5$ d $2n + 1$

7 Solve the inequalities.
Show each solution on a number line.
a $x + 1 < 3$ b $x - 5 \geqslant 12$
c $3x \leqslant 12$ d $\frac{x}{4} \leqslant 7$
e $2x + 1 > 11$ f $3x - 2 > -1$
g $2(x + 5) < 30$ h $3(2x - 4) \geqslant 24$

> **Q7 hint** Use the balancing method – do the same to both sides.

8
a x is an integer.
$-2 \leqslant x < 2$
List the possible values of x. **(2 marks)**
b Write down the inequality shown in the diagram.

```
   -5 -4 -3 -2 -1  0  1  2  3  4  5
```
→ x

(2 marks)
c Solve $2x + 4 > 19$ **(2 marks)**

9 Solve
a $2x + 5 > 4x - 7$ b $3x - 10 \leqslant 5x + 2$
c $10 - 3x \geqslant 4x - 4$ d $10 - x > x - 10$
e $5 - 3x < 10 - 2x$ f $2x + 1 \geqslant 1 - 2x$

10
The sum of a number and 1 more than double the number is more than 5.
What is the smallest positive integer that the number could be? **(4 marks)**

11 **R / P** Tyrone rounds a 3-digit whole number to the nearest 100. He gets 600.
Write an inequality to show what the number could have been.

12 **P** I think of a number, double it and add 4. My answer is greater than when I multiply the number by 3 and subtract 5.
Find three possible values for my number.

5.5 More inequalities

1 Solve
a $4 < 2x \leqslant 16$ b $-3 < 3x \leqslant 1$
c $0 \leqslant 4x < 20$ d $-2 \leqslant 3x \leqslant 15$
e $-7 \leqslant 4x < 10$ f $-4 < 5x \leqslant 18$

2 Solve these double inequalities.
a $11 < 4x - 1 < 23$
b $5 < 2x + 3 \leqslant 21$
c $3 \leqslant 5x - 2 \leqslant 13$
d $-4 < 3x + 8 < 11$
e $-9 < 2x + 1 \leqslant -1$
f $-12 \leqslant 4x + 4 < 0$

Example

3 Solve each two-sided inequality and show each solution set on a number line.
  a $10 < 5x < 25$
  b $14 \leqslant 4x - 2 < 30$
  c $-2 \leqslant 3x \leqslant 2$
  d $4 < 2x + 1 \leqslant 10$
  e $12 \leqslant 3(x + 5) < 21$
  f $-6 \leqslant 3(2x + 3) \leqslant 9$

4 On a number line you can see that $-4 < -3$.

$$\begin{array}{ccccccc} \mid & \mid & \mid & \mid & \mid & \mid & \mid \\ -4 & -3 & -2 & -1 & 0 & 1 & 2 \end{array}$$

  a Multiply both sides of the inequality by -1. Is the inequality still true?
  b $5 < -x$. Write an inequality for x.

5 Solve these double inequalities and show each solution set on a number line.
  a $-6 < -2(x + 3) < 10$
  b $4 < -3x - 2 \leqslant 7$
  c $-2 < 2(5 - 2x) \leqslant 10$

6 Solve the inequality $10 - 5x \geqslant 5 - 2x$
  What is the smallest integer that satisfies it?

7 **P** The sum of three consecutive whole numbers is less than 33.
  What are the largest whole numbers that they could be?

8 **P** The perimeter of the equilateral triangle is more than the perimeter of the square.

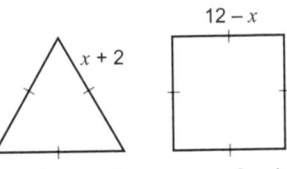

12 − x

x + 2

  Work out the range of values for the side length of the square.

9 For each pair find the integer values of x which satisfy both inequalities.
  a $x > 3$ and $x < 7$
  b $x > -2$ and $x \leqslant 4$
  c $x \geqslant -10$ and $x \leqslant -2$
  d $x > -3$ and $x \geqslant 1$

10 **Exam-style question**

  Find the integer value of x which satisfies both the inequalities
  $x - 4 > 2$ and $3x + 6 < 30$ **(2 marks)**

  Q10 hint Solve each inequality and then show your solutions on a number line to get the final answer.

5.6 Using formulae

1 $d = st$
  a Find the value of t when
    i $d = 20, s = 5$ ii $d = 35, s = 7$
  b Find the value of s when
    i $d = 12, t = 2$ ii $d = 121, t = 11$

  Q1a i hint $d = s \times t$
  $20 = 5 \times t$
  $\square = t$

2 $y = 2x + 5$
  Work out the value of x when
  a $y = 11$ b $y = -1$
  c $y = 105$ d $y = -15$

3 $A = \frac{1}{2}(a + b)h$
  a Work out the value of a when $A = 28$, $b = 4$ and $h = 8$
  b Work out the value of b when $A = 3$, $a = 2$ and $h = \frac{1}{2}$
  c Work out the value of h when $A = 120$, $a = 6$ and $b = 12$

4 $a = \dfrac{b}{3}$
  Work out the value of b when
  a $a = 5$ b $a = \frac{1}{3}$
  c $a = -2$ d $a = 2.1$
  e $a = \frac{5}{6}$ f $a = \frac{2}{3}$

5 $M = 2n + 5q$
  a Work out the value of n when
    i $M = 10$ and $q = 2$
    ii $M = 31$ and $q = 5$
  b Work out the value of q when
    i $M = 10.5$ and $n = 4$
    ii $M = -1$ and $n = 2$

6 The formula to work out the density of an object is $d = \dfrac{m}{V}$ where d = density (g/cm³), m = mass (g) and V = volume (cm³).
  a Work out the mass when
    i $d = 6\,\text{g/cm}^3$ and $V = 5\,\text{cm}^3$
    ii $d = 10\,\text{g/cm}^3$ and $V = 1.5\,\text{cm}^3$.
  b Work out the volume when
    i $d = 6\,\text{g/cm}^3$ and $m = 36\,\text{g}$
    ii $d = 3.6\,\text{g/cm}^3$ and $m = 18\,\text{g}$.

8 In the formula $A = 12r^2$, A is the approximate surface area of a sphere (cm²) and r is the radius (cm).

Use $A = 12r^2$ to work out r when $A = 108\,\text{cm}^2$.

9 You can use this formula to work out the final velocity of an object, v (m/s):

$v^2 = u^2 + 2as$, where u = initial velocity (m/s), a = acceleration (m/s²) and s = distance (m).
Work out the value of a when $v = 6\,\text{m/s}$, $u = 4\,\text{m/s}$ and $s = 5\,\text{m}$.

10 The formula for calculating speed is $s = \dfrac{d}{t}$ where s = speed (m/s), d = distance (m) and t = time (s).

Work out the distance travelled (in metres) by a person running at 9.4 m/s for 130 s.

11 State whether each of these is an expression, an equation or a formula.

a $y = 3x + 2$ b $s = ut + \frac{1}{2}at^2$

c $2x - 5 + y$ d $12 - 3x^2 = 4$

e $s = \left(\dfrac{u + v}{2}\right) t$ f $3(2p + 7)$

12 Rearrange each formula to make the letter in the square brackets the subject.

Example

a $y = 3x$ $[x]$

b $2a = 3b$ $[b]$

c $d = st$ $[t]$

d $y = x + 5$ $[x]$ e $s = 3 + 2t$ $[t]$

f $10t - 3y = 0$ $[y]$ g $a = \dfrac{b}{c}$ $[b]$

h $p = \dfrac{x}{r}$ $[r]$ i $y = \dfrac{2}{3x}$ $[x]$

5.7 Generating sequences

1 Write down the next two terms in each sequence.

a 0.4, 0.8, 1.2, 1.6, ☐, ☐

b $-1, -\frac{1}{2}, 0, \frac{1}{2}$, ☐, ☐

c 1.3, 1.1, 0.9, 0.7, ☐, ☐

d 3.5, 2.4, 1.3, 0.2, ☐, ☐

e $\frac{7}{8}, \frac{5}{8}, \frac{3}{8}, \frac{1}{8}$, ☐, ☐

f −11.5, −11.1, −10.7, −10.3, ☐, ☐

2 Use the first term and the term-to-term rule to generate the first five terms of each sequence.

a start at 5 and add 0.7

b start at 4 and subtract 0.6

c start at 9 and subtract 4

d start at −2 and add 7

e start at −8 and add 3

f start at −3 and subtract 4

3 In a Fibonacci sequence, the term-to-term rule is 'add the two previous terms to get the next one.' Write the next three terms in each Fibonacci sequence.

a 1, 2, 3, 5, 8, …

b 2, 5, 7, 12, 19, …

c 4, 4, 8, 12, 20, …

4 Each sequence is made up of a pattern of dots. For each sequence

i Draw the next two patterns.

ii Draw a table like this:

Pattern number	1	2	3	
Number of dots				

iii Write down the rule to continue the pattern and complete it for each pattern.

iv Work out the number of dots needed for the 10th pattern.

a

b

c

d

5 Write down the next two terms in each sequence.

a 0.1, 1, 10, 100, ☐, ☐   b $3, 1, \frac{1}{3}$, ☐, ☐

c 1, 5, 25, 125, ☐, ☐   d 0.1, 0.01, 0.001, ☐, ☐

6 Find the term-to-term rule for each sequence.
  a 200, 100, 50, 25, …
  b −10, 10, −10, 10, −10, …
  c 1, 2, 4, 8, 16, …
  d 0.2, 2, 20, 200, …
  e 90, 30, 10, $3\frac{1}{3}$, …
  f 3000, 300, 30, 3, 0.3, …

7 This sequence made from counters shows the first three square numbers.

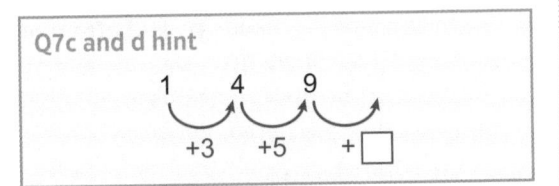

  a Draw the next two patterns in the sequence.
  b Write down the number of counters in each pattern.
  c Work out the differences between the numbers of counters.
  d Follow the sequence to work out the number of counters in the 10th pattern.

  Q7c and d hint
  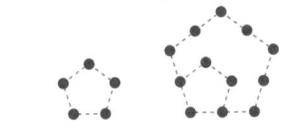

8 This sequence made from counters shows the first three pentagonal numbers.

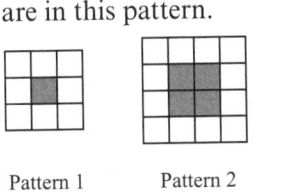

  a Draw the next two patterns in the sequence.
  b Work out the number of counters in the 7th pattern.

9 **Exam-style question**

  Here are some patterns made from grey tiles and white tiles.

  a Draw Pattern 4.
  b Find the number of grey tiles in Pattern 6.

  A pattern has 100 grey tiles.

  c Work out how many white tiles there are in this pattern.

  Pattern 1   Pattern 2   Pattern 3
  (4 marks)

5.8 Using the nth term of a sequence

1 A sequence has nth term $3n − 1$. Copy and complete this table to work out the first five terms of the sequence.

n	1	2	3	4	5
Term	$3 \times 1 − 1 = \Box$	$3 \times 2 − 1 = \Box$			

2 Write the first five terms of the sequence with nth term
  a $2n$ b $n + 5$ c $2n − 5$
  d $10 − n$ e $\frac{1}{2}n + 1$ f $20 − 4n$
  g $−n + 1$ h $−2n + 7$

3 Find the nth term for each sequence.

  Example

  a 3, 7, 11, 15, 19, …
  b 12, 22, 32, 42, …
  c 5, 6, 7, 8, 9, …
  d 10, 8, 6, 4, 2, …
  e 22, 19, 16, 13, 10 …
  f 13, 11, 9, 7, 5, …

4 For each sequence, explain whether each number in the brackets is a term in the sequence or not.
  a 4, 7, 10, 13, …   (34, 61)
  b 3, 8, 13, 18, 23, …   (43, 55)
  c 10, 18, 26, 34, 42, …   (71, 84)
  d −3, 4, 11, 18, 25, …   (70, 100)
  e 40, 32, 24, 16, 8, …   (−16, −40)
  f 20, 17, 14, 11, 8, …   (−2, −10)

  Q4a hint Work out the nth term
  $\Box n + \Box = 34$
  $n = \Box$

5 Using the nth term given, find the 20th term.
  a $5n$ b $2n − 7$ c $12 − 4n$

6 Find the nth term for each sequence. Use it to work out the 10th term.
  a 0, 2, 4, 6, 8, …   b 5, 11, 17, 23, …
  c 12, 11, 10, 9, …   d 6, 2, −2, −6, …

7 Find the first term over 100 for each sequence.
  a 3, 6, 9, 12, …   b 10, 21, 32, 43, …
  c 8, 12, 16, 20, …   d 15, 30, 45, 60, …

8 Here are the first four terms in a number sequence.

37, 33, 29, 25, …

  a Write down the next term in this number sequence.

  b Work out the nth term for this number sequence.

  c Can 2 be a term in this sequence? Give a reason for your answer.

9 Write down the first five terms of the sequence with nth term

  a n^2 **b** $n^2 + 3$ **c** $4n^2$

  d $\frac{1}{2}n^2$ **e** $10 - n^2$ **f** $-n^2$

10 Exam-style question

Here is a pattern made from sticks.

  5   9   13

  a Draw the next pattern in the sequence.

  b Copy and complete this table for the numbers of sticks used to make the patterns.

Term	1	2	3	4	5
Number of sticks					

  c Write, in terms of n, the number of sticks needed for pattern n.

  d How many sticks are needed for pattern 20?   **(5 marks)**

11 P Sarah makes a pattern sequence with squares.

  a Draw the next pattern in the sequence.

  b Sarah has 38 squares.
    Does she have enough to make
    i the 10th pattern
    ii the 20th pattern?

  c Work out the pattern number of the biggest pattern she can make.

12 P Here is a pattern sequence of yellow and green tiles.

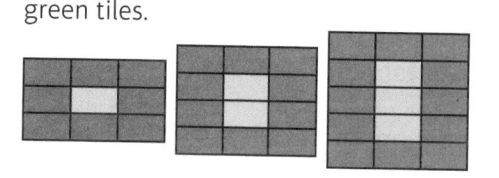

  a Copy and complete the table for the numbers of yellow tiles and the numbers of green tiles.

Pattern number	1	2	3	4	5
Number of yellow tiles	1	2			
Number of green tiles	8	10			

  b Write down the nth term for the sequence of the numbers of yellow tiles.

  c Write down the nth term for the sequence of the numbers of green tiles.

  d How many yellow tiles are there in the 20th pattern?

  e How many green tiles are there in the 30th pattern?

  f Fadeelah has 60 green tiles and 15 yellow tiles.
    Which is the largest complete pattern she can make?

5 Problem-solving

Solve problems using these strategies where appropriate:
- **Use pictures**
- **Use smaller numbers**
- **Use bar models.**

1 Exam-style question

The pie chart below shows the percentages the four nominees received in a class vote.

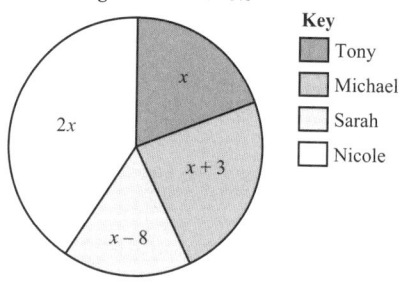

Percentage of the class vote

Key
- Tony
- Michael
- Sarah
- Nicole

The percentages are

x
$x + 3$
$x - 8$
$2x$

What percentage of the vote did Sarah get?
          (4 marks)

2 A rectangle has a height of p cm and a width of $(p + 2)$ cm. The area of the rectangle is 24 cm².
  a Write an equation for the area, A, of the rectangle in terms of p.
  b p is a whole number.
    Solve the equation to find p.

3 **R** Weight is the force exerted on a mass. On Earth, an object's weight (in N) is $10m$, where m is the mass (in kg).
  To calculate the pressure, P (in N/m²), divide the force applied, F, by the area, A.
  a Write down the formula for calculating the pressure in terms of m and A.
  b An elephant's mass is 5000 kg. Each foot has an area of 0.1 m².
    Work out the pressure for each foot.
  c The mass of a woman wearing stiletto heels is 63 kg. Each heel has an area of 0.000 04 m².
    Work out the pressure for each heel.
  d Which will do more damage to a lawn, the elephant or the woman?

4 **R** Ken has £45. He gets £12 a week pocket money and will save half of it each week.
  a Write down how much he will have after the 1st, 2nd, 3rd and 4th weeks.
  b This forms a sequence. How can you tell?
  c What is the nth term of this sequence?
  d How much will he have after 6 months?

> **Q4 hint** Draw bar models to show the amount each week.

5 **R** Hanna bought a new car for £8250. She paid a deposit of £2150 and paid the rest back during the first year.
  From January to May she paid £x each month. She then increased her payments by £100 and paid the new amount for the rest of the year.
  a Show that $12x + 2850 = 8250$
  b How much did Hanna pay back in November?

6 Annie's mum is 165 cm tall. Annie is still not as tall as her mum when she stands on a 20 cm step.
  Use h to represent Annie's height.
  a Write an inequality for Annie's height.
  Annie is more than twice the height of her baby brother. He is 65 cm tall.

  b Write another inequality for Annie's height.
  c Draw on a number line the range of possible values for Annie's height.

7 Three tanks, A, B and C, each have x litres of a solution in them.
  Water is added to each tank to dilute the solution. Tank A now has 460 000 litres more, B has double the original amount and C has 4 times the original amount plus 100 000 litres.
  Tank A has more solution in it than tank B but less than tank C.
  a Work out the possible range of values for x. Write your answer as a double inequality.
  b Write three possible values for x that are multiples of 100 000.

8 **R** A new school is opening with only Year 7 at first. It estimates that the total population of the students and staff will grow like a sequence with an nth term of $125n + 5$ for 7 years. The population will then stay constant.
  a What is the total population of the school in each of the first 4 years expected to be?
  b Will there be a time when the school has a population of 725 when full? Explain your answer.
  c What will be the total population of the school 10 years after opening?

9 **P** Shapes A and B are rectangles. Shape A has a larger area than shape B.

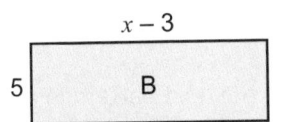

  Work out a range of values for the perimeter of shape A.

10 **R** A tank is 1 m deep. David fills the tank with water. The water is 12.4 cm high when he starts. The water level in the tank rises by 5.6 cm every minute.
  a What is the water level after 3 minutes?
  b What is the water level after n minutes?
  c The tank must not be completely filled. Write an inequality for the water level in terms of n.
  d Solve your inequality. What does it tell you?

6 ANGLES

6.1 Properties of shapes

1   Write down the letters of two pairs of shapes that are congruent.

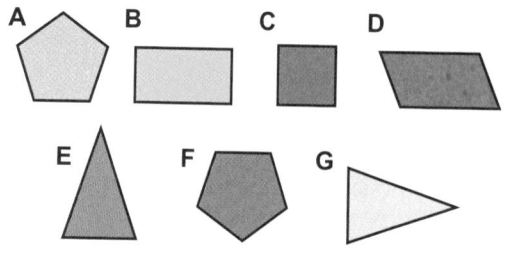

> **Q1 hint** Congruent shapes are exactly the same size and shape, so if you cut them out, one shape fits on top of the other exactly. It does not matter if you turn them over or turn them round.

2   These parts of shapes are drawn on a centimetre square grid. The dotted lines are lines of symmetry.

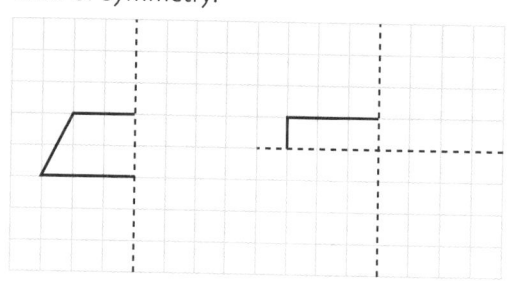

   a Copy and complete the diagrams.
   b Name each shape and measure and label all its angles.

3   Which of these shapes have
   a equal length diagonals
   b two pairs of equal sides
   c only one pair of parallel sides?

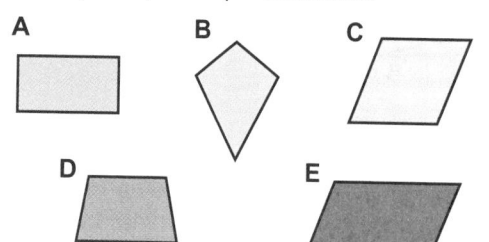

4   **R** Name each quadrilateral being described.
   a I have two pairs of parallel sides and my opposite angles are equal.
   b I have four equal sides and my diagonals are not the same length.
   c I have two pairs of equal sides and one pair of equal angles.
   d My diagonals are the same length and bisect at 90°.

5   **P** Draw a coordinate grid with axes labelled from −5 to 5.
   Plot these points.
      A(−1, 0), B(1, −1), C(1, 5)
   Write down the coordinates of a point that would make
   a a kite
   b a parallelogram
   c an isosceles trapezium.

6   Work out the sizes of the missing angles in each quadrilateral.

   a 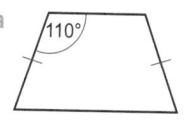   b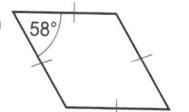

   c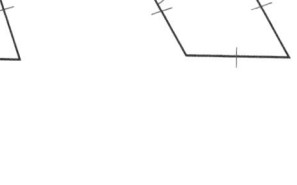

   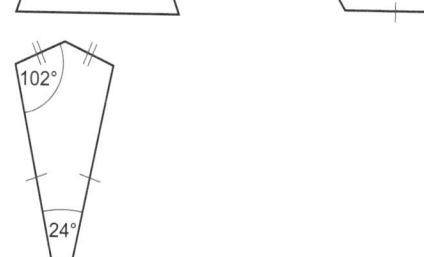

7   **R** Charlie is an architect.
   He draws a design for a path.
   The left-hand section is a square and the others are rhombuses.

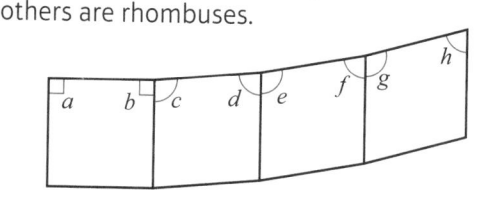

   Charlie uses these rules in his design.
   • Angle c is 5% bigger than angle a, rounded to the nearest degree.
   • Angle e is 5% bigger than angle c, rounded to the nearest degree.
   • Angle g is 5% bigger than angle e, rounded to the nearest degree.
   Work out the sizes of the angles a to h.

8 **Exam-style question**

ABCD is a quadrilateral. Find the size of angle BCD.

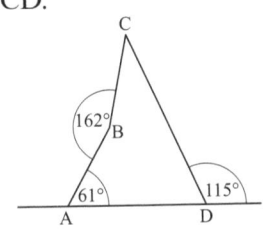

Explain each stage of your working.

(3 marks)

9 **P** Draw a coordinate grid with axes labelled from −5 to 5.

a Here are some coordinates for three shapes ABCD. Plot the points on your grid.

  i A(1, −2), B(−3, −2), C(−3, 1)

  ii A(0, −2), B(1, 2), C(0, 4)

  iii A(−1, 1), B(2, −1), C(5, 1)

b Join the points A to B and B to C in each shape.

c Draw a point D for each shape so that it has rotation symmetry of order 2.

d Name each shape.

10 **P** The diagram shows a hexagon made from three identical rhombuses.

Show that angles a, b and c add up to 360°.

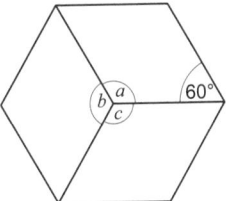

11 **P** The diagram shows a kite.

Find the size of angle a.

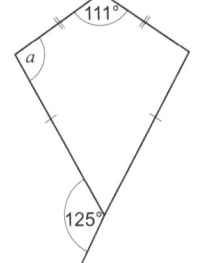

6.2 Angles in parallel lines

1 The diagram shows a line crossing two parallel lines and angles labelled f, g, h and i.

Write down two pairs of alternate angles.

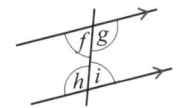

2 Find the sizes of the angles marked with letters.

a

b

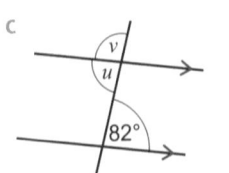

c

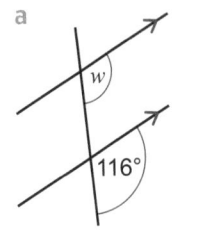

3 Find the sizes of the angles marked with letters.

a

b

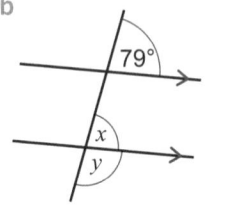

4 Find the sizes of angles a to e.

Give reasons for your answers.

Choose your reasons from the box.

> Angles on a straight line add up to 180°.
> Angles ☐ and ☐ are alternate and equal.
> Angles ☐ and ☐ are corresponding and equal.

a

b

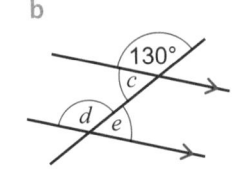

5 Which angles in the diagram are equal?

Give reasons for your answers.

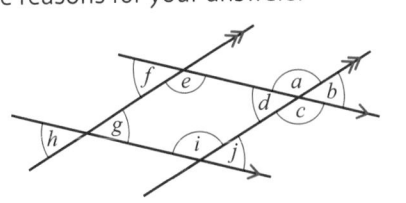

6 Angles inside two parallel lines are called **co-interior angles**. Use this diagram to explain why co-interior angles add to 180°.

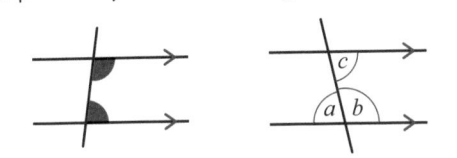

7

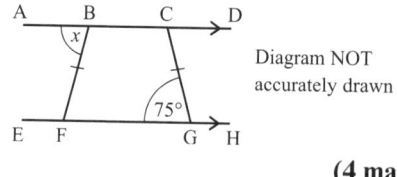

Exam-style question

ABCD and EFGH are straight lines.
BCGF is an isosceles trapezium.
Find the size of the angle marked *x*.
Give reasons to explain your answer.

Diagram NOT
accurately drawn

(4 marks)

8 Find the sizes of the angles *a* to *j*.
Give reasons for your answers.

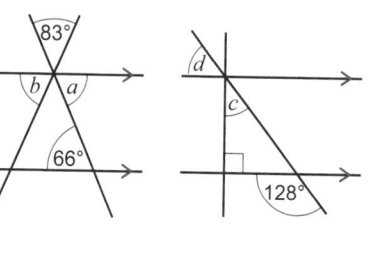

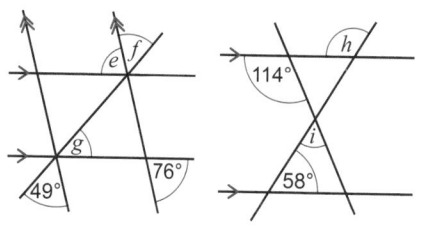

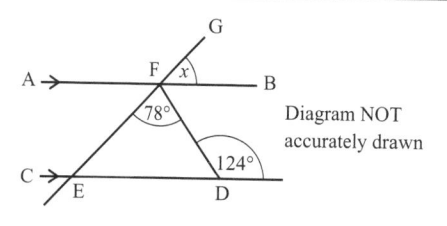

9

Exam-style question

AFB and CED are parallel lines.
EFG is a straight line.
Work out the size of the angle marked *x*.

Diagram NOT
accurately drawn

(3 marks)

10 PQRS is a trapezium.
Work out the sizes of the angles *a* and *b*.

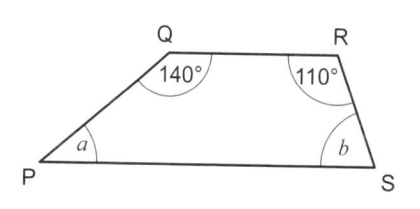

6.3 Angles in triangles

1 Calculate the sizes of the angles marked
with letters.

a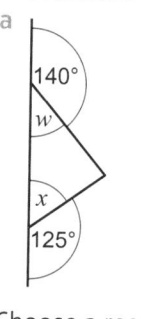

b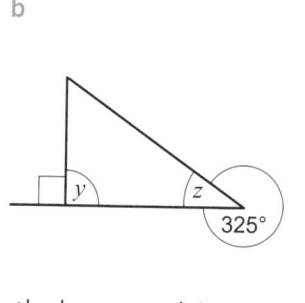

Choose a reason from the box to explain
each answer.

> Reason 1 Angles about a point add up to 360°.
> Reason 2 Angles on a straight line add up
> to 180°.

2 **R** The diagram shows an isosceles triangle
and a scalene triangle.

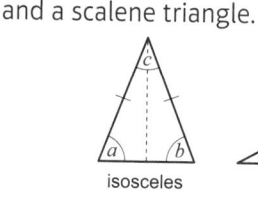

isosceles   scalene

a An isosceles triangle has a line of symmetry.
What does this tell you about the two
angles *a* and *c*?

b A scalene triangle has no lines of symmetry.
What does this tell you about angles *d*, *e*
and *f*?

3 Find the sizes of the missing angles.

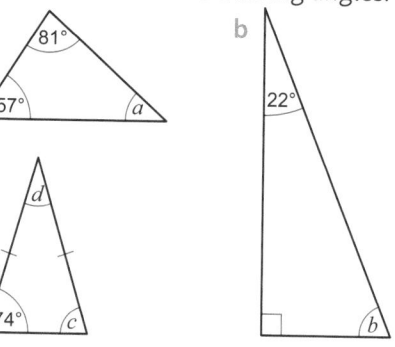

4 In order to be usable, a ramp should make an
angle of 20° or less with the horizontal.
a How many degrees is this with the vertical?
b Is either of these ramps usable? Explain
your answer.

5 An engineer designs this support for a beam. What sizes are angles a and b?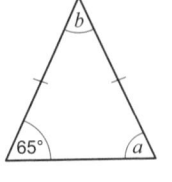

6 **R** Sketch a copy of this diagram. Copy and complete these statements to prove that the sum of the angles in a triangle is 180°.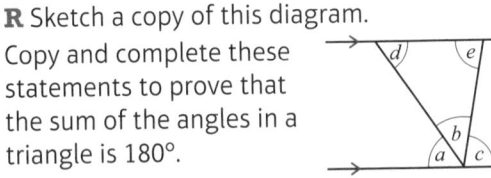

Statement	Reason
$a + b + c = \square$	Angles a, b and c lie on a straight line.
$d = \square$	Angles d and $\square$ are alternate.
$e = \square$	Angles e and $\square$ are alternate.

So $d + b + e = a + b + c$
$= \square$

This proves that the angles in a triangle sum to $\square$.

7 **P** Work out the sizes of the angles marked with letters.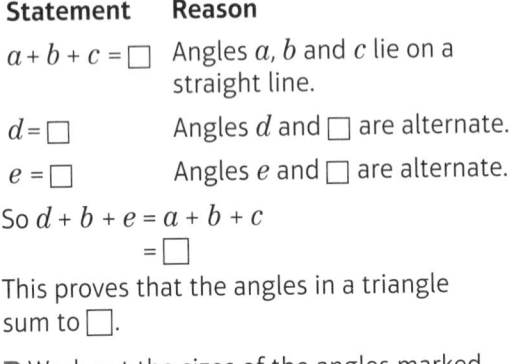

a   b

8 **R** Copy the diagram.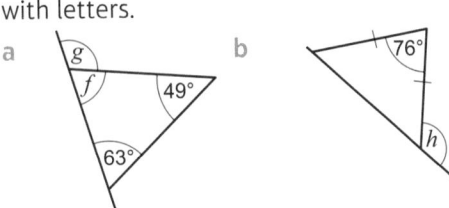

Copy and complete this proof to show that $s = p + q$.

$p + q = 180° - \square$ because the angles in a triangle add up to $\square$.

$s = 180° - \square$ because r and s lie on a _____ line.

So $p + q = s$

9 **R** Which statement is true about this triangle?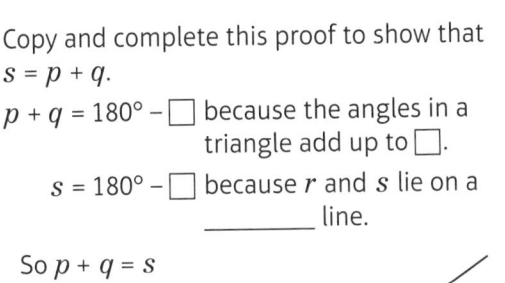

A Angle z = angle y
B Angle z = angle w + angle x
C Angle z = angle x + angle y
D Angle z = angle w + angle y

10 **R** What sizes could angles a, b and c be? Choose possible angles from the cloud.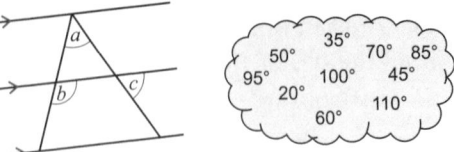

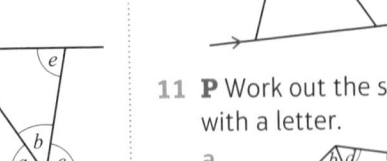
50° 35° 70° 85°
95° 100° 45°
20° 110°
60°

11 **P** Work out the sizes of the angles marked with a letter.

a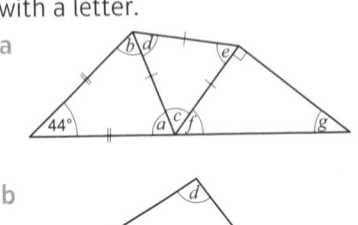

b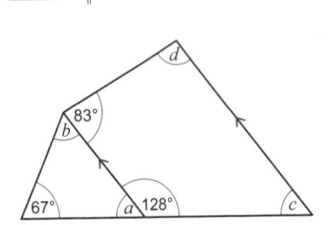

Q11a hint What types of triangles can you see?
Q11b hint What do the angles of a quadrilateral add up to? How can you use properties of parallel lines?

12 **Exam-style question**

AB is parallel to CD.
CF = EF
$\widehat{CFE} = 72°$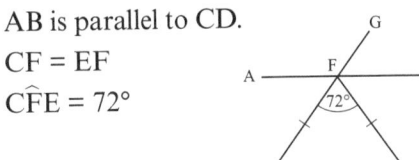

Calculate the size of $\widehat{AFG}$. **(3 marks)**

6.4 Exterior and interior angles

1 Copy the polygons. Draw one interior angle and one exterior angle for each shape.
a   b   c   d
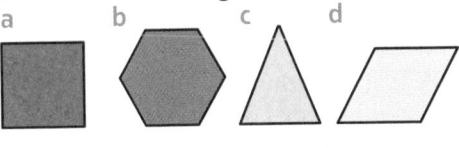

2 Write down whether each polygon is regular or irregular.
a   b   c   d
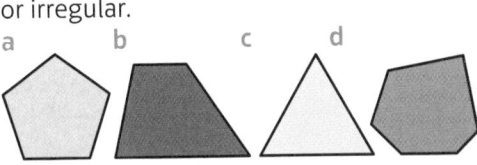

3 **a** Work out the size of one exterior angle of each regular polygon.

i   ii

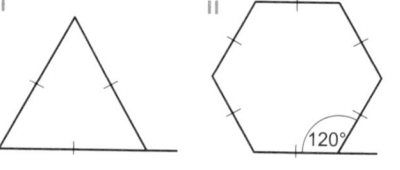

b Work out the sum of the exterior angles of each regular polygon. What do you notice?

> **Q3 hint** In a regular polygon, all the exterior angles are equal.

4 A regular polygon has 12 sides.
  a How many equal exterior angles does it have?
  b Work out the size of *one* exterior angle.

5 The exterior angles of some regular polygons are
  a 20°   **b** 24°   **c** 36°
  Work out the number of sides of each regular polygon.

6 The diagram shows an irregular hexagon. What do the exterior angles add up to?

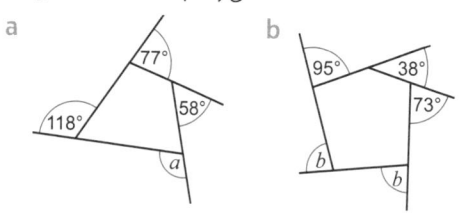

7 Work out the sizes of the missing exterior angles for each polygon.

a   b

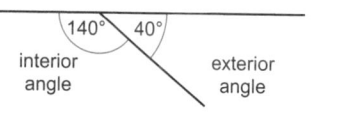

8 **R** Anna says, 'I can work out the size of the exterior angles of an irregular octagon by dividing 360° by 8.'
  Is Anna correct? Explain.

9 **R** The diagram shows an exterior and an interior angle of a regular polygon.

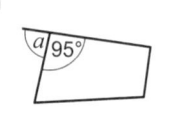

a Work out the number of sides of the polygon.
b What do the interior and exterior angles add up to?
c Copy and complete this rule.
  The interior and exterior angles always sum to ☐°.
d Explain why this rule is true.

10 Work out the sizes of the angles marked with letters.

a   b

11 Copy and complete this table.

Regular polygon	Exterior angle	Interior angle
triangle		
hexagon		
octagon		

12 A regular polygon has an interior angle of 168°. Find the size of the exterior angle and work out how many sides the polygon has.

13 **P** Point J lies on the midpoint of EF. Find the size of angle x.

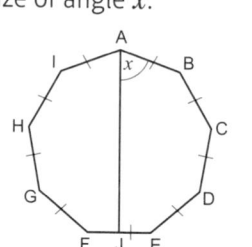

14
Exam-style question

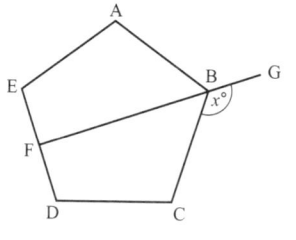

ABCDE is a regular pentagon.
FBG is a line of symmetry.
Angle CBG = x°
Work out the value of x.   **(4 marks)**

6.5 More exterior and interior angles

1 Copy this hexagon onto triangular paper.

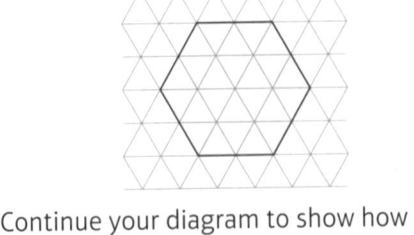

Continue your diagram to show how hexagons fit together.
Show that hexagons fit together by considering the interior angles.

2 Do regular heptagons fit together? Explain your answer by considering the interior angles.

3 Copy this triangle onto squared paper.

Show how triangles fit together by drawing five more on your diagram.

4 **R** Copy and complete the statements below.

This shape has ☐ sides.
It is made of ☐ triangles.
The interior angles in a triangle add up to ☐°.
So the sum of the interior angles in this shape is ☐ × ☐° = ☐°

5 a **R** What is the sum of the interior angles of this polygon?

b How many triangles can you divide an n-sided polygon into?

c Write a formula to work out the sum of the interior angles, S, for a polygon with n sides.

6 For each irregular polygon, work out
i the sum of the interior angles
ii the size of the angle marked with a letter.

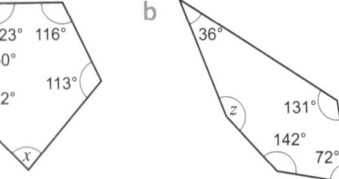

7 For each shape, work out the size of
i the angle sum
ii the interior angle.

a regular octagon
b regular nonagon
c 15-sided regular polygon

8 For each polygon, work out the number of sides from the sum of its interior angles.
a 900°   b 1980°
c 2880°   d 5220°

9 The diagrams show regular polygons with some angles given.

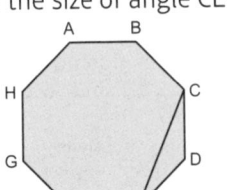

Work out the sizes of the angles marked with letters.
Give reasons for your answers.
Choose reasons from the box.

> The sum of the interior angles of a regular polygon = $(n - 2) \times 180°$.
> The triangle is isosceles.
> Exterior and interior angles add up to 180°.

Q9b hint How many sides does this regular polygon have?

10 **P** ABCDEFGH is a regular octagon.
Find the size of angle CED.

Example

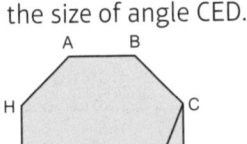

11

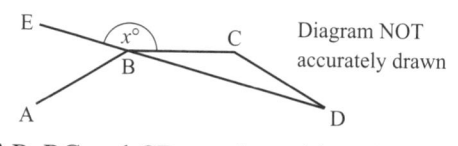

AB, BC and CD are three sides of a regular 12-sided polygon.

EBD is a straight line.

Angle EBC = $x°$

Work out the value of x. **(4 marks)**

6.6 Geometrical problems

1 Work out the value of x in each diagram.

Example

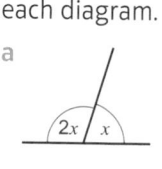 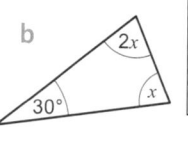

a   b

c

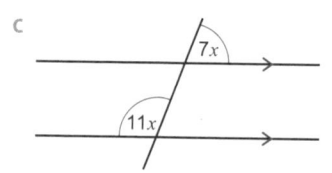

2 **R** Look at the diagram.

a   i   Explain why $x + 50° = 2x + 30°$

    ii   Work out the value of x.

    iii   Work out the sizes of the angles.

b   Work out the sizes of the angles.

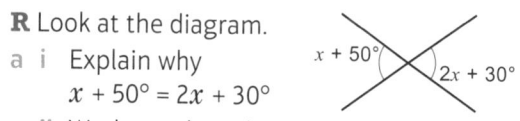

3 **P** Work out the values of a and b.

a

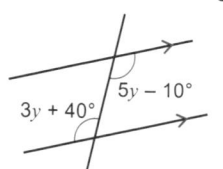

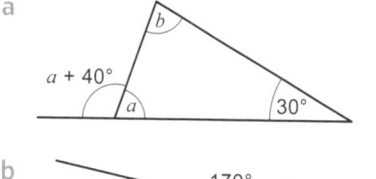

b

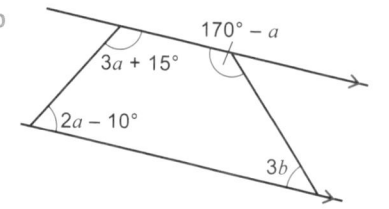

4 Look at the diagram.

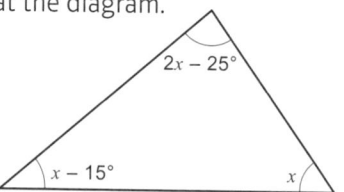

a   Write an equation in terms of x.

b   Solve this equation for x.

c   Write down the sizes of the three angles of the triangle.

5 Write an equation in terms of x and use it to find the three angles of the triangle.

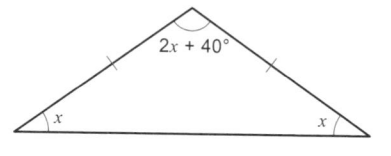

6 a   The left-hand column of the table shows the angles in three different triangles.

The right-hand column shows equations that have been written about the triangles, but not in order.

Angles	Equations
50°, 60°, 70°	$3a = 180°$
20°, 70°, 90°	$2a + b = 180°$
60°, 60°, 60°	$a + b = 90°$
25°, 25°, 130°	$a + b + c = 180°$

Match each set of angles with the correct equation.

b   Here are three more angles: 40°, 50°, 80°.

Explain why these three angles do not come from a triangle.

7 **R** What type of triangle is this? Explain.

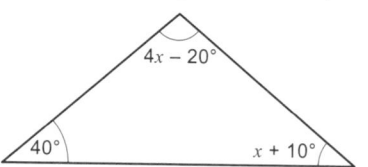

8 In the box are sets of three angles, written as expressions in x.

$2x$, $4x$, $4x$	$3x$, $2x$, x
$100° - x$, $x + 60°$, $20°$	
$x + 50°$, $x - 50°$, $x + 120°$	
$2x + 30°$, $50° - x$, $40° - x$	

For each set, which of these is true?

These three angles can always/sometimes/never make a triangle.

Exam-style question

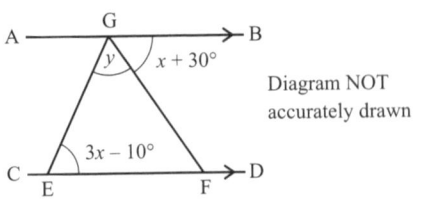

All angles are measured in degrees.

AGB and CEFD are straight lines.

Angle GEF = $3x - 10$

Angle BGF = $x + 30$

Angle EGF = y

a Show that $y = 160 - 4x$

Give reasons for each stage of your working. **(3 marks)**

b Given that $x = 25$,

  i work out the value of y

  ii work out the size of the largest angle in triangle EGF. **(4 marks)**

Exam hint

'Show that' means you need to write down all of your working and explain your reasons as you go along. You can answer part **b** even if you cannot answer part **a**.

10 P Triangle ACD is an isosceles triangle.
Triangle BDE is a right-angled triangle.
Show that triangle ABD is an equilateral triangle.

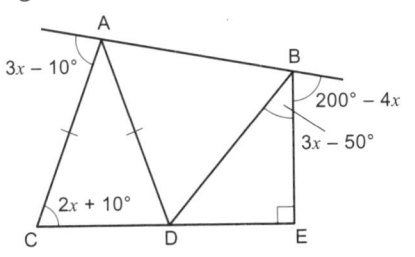

11 P ABC is a triangle.
D is a point on AC.
Find the size of angle ACB.

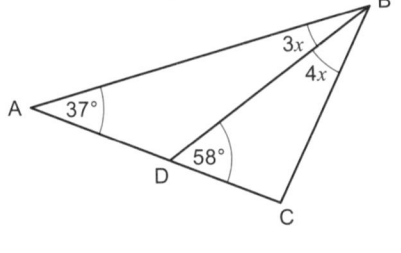

6 Problem-solving

Solve problems using these strategies where appropriate:
- Use pictures
- Use smaller numbers
- Use bar models
- Use x for the unknown.

Example

1 R I think of a 2D shape. It has only four sides. They are all straight.

  a What could my shape be? Write a list.

  It has two different pairs of equal length sides.

  b What could my shape be now? Write a new list.

2 R A regular polygon has an even number of sides and fewer than six sides.

  a What is the name of the shape?

  b What is the size of an interior angle of the polygon?

3 A quadrilateral has two pairs of equal angles. The angles next to each other are different.

  a What type of quadrilateral is this?

  b One of the angles is 65°.
  What are the other three angles?

4

Exam-style question

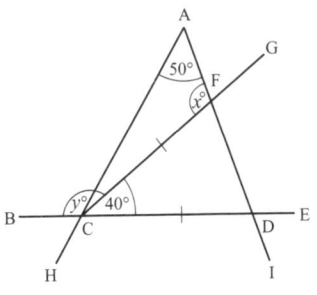

Angle AFC = $x°$ and Angle ACB = $y°$.

a What is the value of x? **(2 marks)**

b What is the value of y? **(2 marks)**

Give reasons for your answers.

5 In the quadrilateral below, x is $\frac{2}{3}$ of $\angle$ABC and y is $\frac{5}{6}$ of $\angle$ABC.

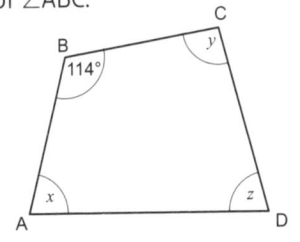

What are the values of x, y and z?

6 AG is parallel to HD.
AB = AF and CBG = 66°.

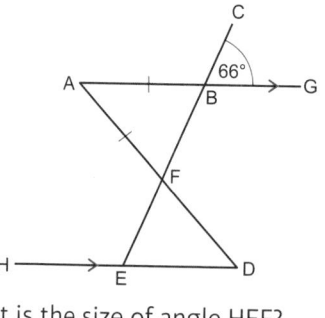

a What is the size of angle HEF?
b Are FD and FE the same length?
How can you tell?

7 Angle ACB is double the size of angle BAC.

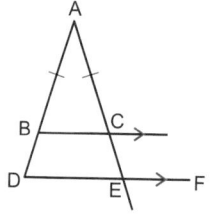

a Write an equation for the sum of the angles in triangle ABC.
b What is the size of ∠ABC?
c What is the size of ∠CEF?
Give reasons for your answers.

Q7 hint Use x for the unknown.

8 **R** Karl says, 'Angles a, b and c must add up to 180° in the polygon PQRST.'

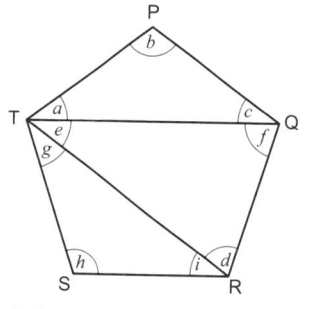

a Is he right? How can you tell?
b Use the diagram to show that the sum of the interior angles in a pentagon add up to 540°.

9 The diagram below shows part of a regular polygon and an extended line.
The exterior angle is one-fifth of the size of the interior angle.

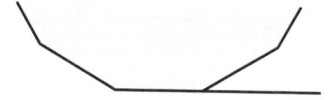

a What is the size of the exterior angle of the polygon?
b How many sides does the polygon have?

10 Exam-style question

The diagram shows two congruent regular heptagons on a straight line.
Angle CAB = $x°$.

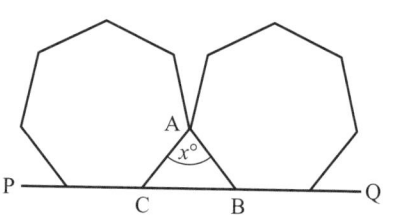

a Show that triangle ABC is **not** an equilateral triangle. **(3 marks)**
b What is the size of x to 1 d.p.? **(2 marks)**

11 Two regular polygons share a common side BE.
EBX is a straight line.
Angle ABC = angle DEF = 90°.
Angle CBX is half of angle ABX.

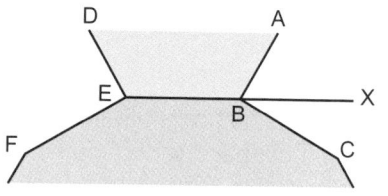

a What is the size of angle ABE?
b Name the polygons.

7 AVERAGES AND RANGE

7.1 Mean and range

1 Here are the ages, in years, of the first 10 customers in a shop.
44, 65, 73, 12, 38, 3, 67, 81, 19, 28
Work out the mean age of the customers.

Example

2 The table shows the rainfall for the first four months of one year.
Work out the mean rainfall per month.

Month	January	February	March	April
Rainfall (mm)	160	97	108	145

3

> **Exam-style question**
>
> The table shows the scores for 10 students in a maths exam.
>
> | Student | A | B | C | D | E | F | G | H | I | J |
> |---|---|---|---|---|---|---|---|---|---|---|
> | Score | 23 | 25 | 20 | 19 | 18 | 19 | 20 | 17 | 16 | 20 |
>
> **a** Find the range of scores.
> **b** Write down the mode.
> **c** Work out the mean score.   **(5 marks)**

4 Mr Adiga records how late his train is each morning, in minutes.
3, 5, 6, 5, 3, 2, 1, 0, 0, 1, 1, 0, 1, 2, 1, 12, 1, 4, 4, 5
a Work out the mean time the train is late.
b Work out the range.

5 Here are the numbers of films that customers using a broadband service buy in a month.
3, 0, 1, 1, 2, 1, 0, 1, 0, 1, 1, 1, 2, 2, 2, 3, 2, 2, 2, 3
The numbers are shown again in the table.

Number of films downloaded, n	Frequency, f	$n \times f$
0	3	$0 \times 3 = 0$
1		
2		
3		
Total		

a Copy and complete the table.
b Find the total number of films downloaded.
c Find the total number of customers asked.
d Use your answers to parts **b** and **c** to work out the mean number of films downloaded.

6 The weights of 100 newborn babies were recorded, to the nearest kg.
The results are shown in the table.

Weight, w (kg)	Frequency, f	$w \times f$
1	3	
2	19	
3	47	
4	30	
5	1	
Total		

a Work out the mean. **b** Work out the range.
Give your answers to an appropriate degree of accuracy.

7 **R** The charts show the times, to the nearest 5 minutes, spent by students in class 11A and class 11B on their maths homework.

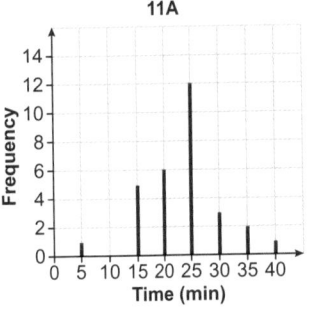
11A

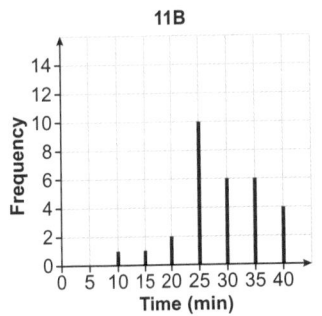
11B

a Copy and complete the frequency table for class 11A.

Time, t (min)	Frequency, f	$t \times f$
5	1	
10	0	
15	5	
20		
25		
30		
35		
40		
Total		

b Work out the mean and range for class 11A. Give your answer correct to 1 decimal place.
c Make a frequency table for class 11B.
d Work out the mean and range for class 11B.
e Choose words from the box to fill the gaps.
Class .... works harder than class .... because their mean is .... and their range is .... A smaller range shows the data is .... spread out.

higher   more   smaller   11A   less   11B

8 **R** Two football teams recorded how many goals they scored in 10 matches.
Team A: 3, 4, 4, 3, 2, 5, 2, 4, 2, 1
Team B: 0, 1, 7, 1, 0, 5, 1, 8, 6, 1
a Find the mean score for each team.
b Find the range for each team.
c Which team is more consistent?
d Which team would you support? Give a reason.

9 **Exam-style question**

Adele rolls a dice 5 times.
Her first 4 scores are 3, 5, 2 and 6.
The mean of her 5 scores is 4. What does she get on her final roll?   **(3 marks)**

Q9 hint Work out what the total of the 5 rolls of the dice must be.

10 **Exam-style question**

The table shows the numbers of computers owned by 30 families who live on the same road.

Number of computers	Frequency
0	1
1	7
2	12
3	5
4	3
5	2

Work out the mean number of computers per family.
Give your answer correct to 1 decimal place.   **(3 marks)**

7.2 Mode, median and range

1 The stem and leaf diagram shows the prices, to the nearest £1000, of cars in a showroom.

```
0 | 9 9
1 | 1 2 2 3 5 7 8
2 | 9 9
3 | 2 5
4 | 1
```

Key

```
0 | 9  means £9000
```

a How many cars are there?
b Work out the range of the prices.

2 **R** Why is the mode not a useful average for the data in **Q1**?

Example

3 This stem and leaf diagram shows the weights of cats taken to a vet clinic.

```
1 | 3 7 8 9
2 | 0 0 3 5 8 9
3 | 0 2 4
4 | 1 3 5
5 | 2
```

Key

```
2 | 0  means 2.0 kg
```

Find
a the mode   b the median   c the range.

4 **Exam-style question**

Aatami recorded the lengths, in millimetres, of a sample of leaves.
He drew this stem and leaf diagram for his results.

```
3 | 1 3 5 7
4 | 2 2 2 6 7 9
5 | 3 3 9
6 | 0 1
```

Key

```
3 | 1  represents 31 mm
```

a Write down the number of leaves in the sample.   **(1 mark)**
b Write down the mode.   **(1 mark)**
c Work out the range.   **(2 marks)**

5 A maths teacher records students' marks in the end-of-year test.

```
5 | 6  6  7  7  8  9  9

6 | 0  1  4  5  5  7  7  8  9

7 | 2  4  5  5  6  7

8 | 5  6  9  9  9

9 | 8
```

Key

5 | 6 represents 56%

Find

a the mode   b the median   c the range.

> **Q5b hint** There are 28 values. The position of the median is $\dfrac{28 + 1}{2} = 14.5$, so the median is halfway between the 14th and 15th values.

6 **R** For each set of experimental data below
   i identify the outlier
   ii work out the range
   iii work out the range ignoring the outlier.
   a 24°, 18°, 19°, 22°, 12°, 19°, 25°, 17°
   b 3.7 mm, 3.9 mm, 2.9 mm, 4.8 mm, 5.2 mm, 0.4 mm, 5.6 mm, 3.9 mm
   c 1.02 kg, 0.75 kg, 1.09 kg, 0.98 kg, 0.95 kg, 1.03 kg, 1.05 kg

7 Marta recorded these sets of experimental data. The outliers are likely to be errors. Work out the range of each data set, ignoring any outliers.
   a 17, 19, 18, 13, 4, 15
   b 9.2, 8.9, 9.4, 3.5, 8.1, 8.9, 9.5

8 Jayne measured the lengths and masses of a sample of metal rods. The diagram shows the results.

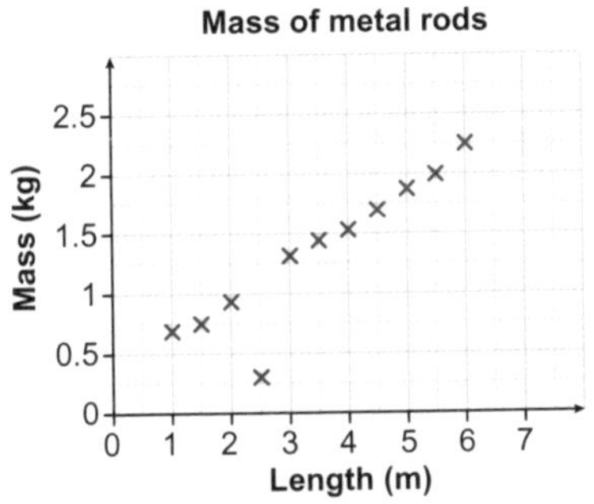

Mass of metal rods

Find the range of the masses, ignoring any outliers.

9 In a traffic survey, the number of cars passing each minute was recorded.

Number of cars	Frequency
0–10	1
11–20	2
21–30	17
31–40	8
41–50	3

Estimate the range of the number of cars.

10 A factory records the weight, in grams, of bags of flour.
The table shows the results.

Weight, w (grams)	Frequency
$960 < w \leqslant 970$	14
$970 < w \leqslant 980$	56
$980 < w \leqslant 990$	125
$990 < w \leqslant 1000$	391
$1000 < w \leqslant 1010$	25

a Write down the largest possible weight that one of these bags of flour could have.
b Write down the smallest possible weight that one of these bags of flour could have.
c Estimate the range.

11 The instructor records the weight (in kg) of participants at a fitness class.

Weight, w (kg)	Frequency
$60 \leqslant w < 65$	14
$65 \leqslant w < 70$	12
$70 \leqslant w < 80$	18
$80 \leqslant w < 90$	3

Estimate the range of the weights.

7.3 Types of average

1 Find the mode of each data set.
   a 54, 59, 55, 56, 52, 56, 51, 56, 59
   b 0.5, 0.9, 0.5, 1.1, 1.3, 1.4, 1.6, 1.2
   c Is the mode a good representative value for both data sets?

2 Find the mean of each data set.
   a 105, 109, 110, 82, 111
   b 3, 4, 4, 2, 4
   c Is the mean a good representative value for both data sets?

3 Find the median of each data set.
   a 0.2, 1.1, 1.5, 1.6, 0.3, 0.3, 0.2
   b 101, 104, 99, 105, 107, 234, 105
   c Is the median a good representative value for both data sets?

4 **R** Work out the most appropriate average for each data set.

a 6, 7, 5, 6, 9, 2, 6, 54

b 3.2, 5.4, 3.2, 3.2, 10.8, 4.5, 4.9

c TV, TV, DVD, film, TV, DVD

d 83, 79, 81, 83, 82

5 a Find the median of 1.2, 3.4, 5.6, 7.6, 8.5

b Find the median of 1.2, 3.4, 5.6, 7.6, 18.5

c Find the median of 1.2, 3.4, 5.6, 7.6, 28.5

6 Decide if each statement is true or false.

a The median will always change if one data value changes.

b With numerical data, there will always be a mean.

c The mean will be affected by extreme values.

d The mode will always be affected by extreme values.

e The mode is the only average that can be used with non-numerical data.

7 The table shows the numbers of library books borrowed by customers one day.

Number of books	Frequency
1–2	15
3–4	12
5–6	5
7–8	2

What is the modal class?

8 This table shows the times taken some students to run 1500 metres.

Time, t (min)	Frequency
$7 < t \leqslant 8$	13
$8 < t \leqslant 9$	19
$9 < t \leqslant 10$	4
$10 < t \leqslant 11$	1

Write down the modal class.

9 Andrew rolled a dice 50 times. The table shows his results.

Example

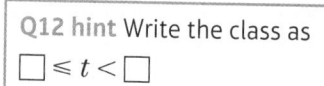

Score	Frequency
1	7
2	9
3	11
4	6
5	8
6	9

Find the median score.

10 Megan recorded the number of lengths she swam each week for 6 months.

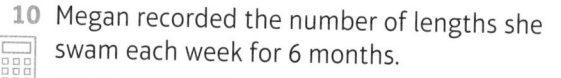

Number of lengths	Frequency
50	12
51	5
52	2
53	7

a Find the median number of lengths she swam.

b What is the mode?

c Work out the mean number of lengths she swam (to 1 d.p.).

11 The number of cars using a bridge during rush hour is recorded over a month.

Number of cars	Frequency
250–300	7
301–350	12
351–400	9
401–450	3

Which class contains the median?

12 **Exam-style question**

This table gives information about the amount of extra time, in minutes, awarded by a football referee.

Extra time, t (min)	Frequency
$0 \leqslant t < 2$	4
$2 \leqslant t < 4$	11
$4 \leqslant t < 6$	3
$6 \leqslant t < 8$	1

a Write down the modal class interval. **(1 mark)**

b Find the class interval that contains the median. **(2 marks)**

c How many matches had 4 or more minutes of extra time? **(2 marks)**

Q12 hint Write the class as
$\square \leqslant t < \square$

7.4 Estimating the mean

1 Write each of these in hours and minutes to the nearest minute.
   a 4.21 hours   b 1.072 hours   c 5.391 hours

2 In a survey, 20 people were asked the distance, in miles, they travel to work.
   This table shows the results.

Example

Distance, d (miles)	Frequency, f	Midpoint of class, m	$m \times f$
1–9	9		
10–18	3		
19–27	7		
28–36	1		
Total		**Total**	

Calculate an estimate for the mean distance travelled.

> **Q2 hint** To find the midpoint of a class add together the endpoints and divide by 2.
> The midpoint of the class 1–9 is $\dfrac{1 + 9}{2} = 5$

3 This table gives information about the time, in minutes, a doctor spends with her patients.

Time, t (min)	Frequency
1–10	8
11–20	23
21–30	14
31–40	5

Work out an estimate of the mean time the doctor spends with each patient.
Give your answer to the nearest minute.

4 The table shows how long, in seconds, some children could hop for.

Time, t (seconds)	Frequency, f
$30 < t \leqslant 60$	2
$60 < t \leqslant 90$	11
$90 < t \leqslant 120$	19
$120 < t \leqslant 150$	3

Work out an estimate of the mean time.
Give your answer to the nearest second.

5 The rainfall (in cm) for a town in the UK is recorded each day during April.
0.3, 0.9, 0, 0, 0.4, 0, 0, 0.5, 0.1, 0.1, 0.2, 0.2, 0.7, 0.2, 0.1, 0.5, 0.3, 0.2, 1.1, 0.8, 0.8, 0.6, 0.3, 0, 0, 0.3, 0.4, 0.7, 0.9, 1.1

a Copy and complete the frequency table.

Rainfall, r (cm)	Frequency, f		
$0 \leqslant r < 0.3$			
$0.3 \leqslant r < 0.6$			
$0.6 \leqslant r < 0.9$			
$0.9 \leqslant r < 1.2$			

b Work out an estimate of the mean rainfall.
c What is the modal class?
d Which class contains the median?
e Estimate the range of the rainfall.

6 **Exam-style question**

This table gives information about the heights, in metres, of a class of students.

Height, h (m)	Frequency, f		
$1.2 \leqslant h < 1.4$	2		
$1.4 \leqslant h < 1.6$	13		
$1.6 \leqslant h < 1.8$	9		
$1.8 \leqslant h < 2$	1		

a Work out an estimate of the mean height of the students. Give your answer to an appropriate level of accuracy. **(5 marks)**
b Explain why your answer is only an estimate. **(1 mark)**

7.5 Sampling

1 **R** A chocolate bar manufacturer wants to know how many people in the UK buy their brand.
   a What is meant by the population in this case?
   b Explain why a sample is needed.
   c The UK population is approximately 64 million. How many people should they choose for the sample?
   100   1000   10 000   100 000   1 000 000

2 Saira wants to find out what percentage of the population in her town walk to their workplace. She carries out a survey at her school.
   a What is meant by the population in this case?
   b Comment on the sample Saira is using.

3 Mahnoor wants to find out how much time people spend watching TV each day. He asks his friends at his cricket club.
   a Explain why Mahnoor's sample is not reliable.
   b How could Mahnoor improve his survey?

4 **R** Michel and Brian are carrying out a survey to find out how long people spend on their Christmas shopping.

Michel chooses a sample at a shopping centre.
Brian chooses a sample at the local gym.
Would you expect them to get similar results?

5 The manager of a swimming pool wants to find out what customers think of the facilities. She decides to ask everyone who comes on a Saturday morning.
a Will this give fair or biased results?
b How can she improve her survey?

6 The owner and the manager of a beauty salon are both selecting a sample of 10 customers.
a The manager selects her sample by picking the first 10 customers who have an appointment that day.
Will this give a random sample?
Explain your answer.
b The owner selects her sample by giving each customer a number and then using a computer program to randomly generate a number to pick.
Will this give a random sample?
Explain your answer.

7 a Here is a list of random numbers.
234, 892, 481, 363, 945, 651, 699, 152, 301, 392
i Write down two digits from each number.
ii Explain how you could use the two digits to pick 10 people at random from a numbered list.
b Mrs Avery has a list of 100 students. She wants to pick 10 students at random to ask about school lunches. Explain how she could use the random numbers from part a to pick students from the list.

Q7a hint 2(34) (89)2

8 Exam-style question

A gym wishes to improve its customer service. It has 250 registered users. The manager wants to carry out a survey using a random sample.

Describe two ways she could select a random sample. **(2 marks)**

9 **R** An online survey is used to predict people's spending on foreign holidays next year.
The company surveyed 6400 people. They picked 4000 from those registered online with an email address and picked the remaining 2400 from the electoral role.

From the 6400 people who responded, the average spend on a foreign holiday the following year was predicted to be £850 per person.
The actual spend per person that year was approximately £950 per person.
a Was the survey accurate?
b Explain why the sample might be biased.

7 Problem-solving

Solve problems using these strategies where appropriate:
• **Use pictures**
• **Use smaller numbers**
• **Use bar models**
• **Use x for the unknown.**

1 Exam-style question

The pictogram shows the number of calculators sold by the maths department each month for five months.

September	🖩 🖩 🖩
October	🖩 🖩
November	🖩 🖩 🖩
December	
January	

Key
🖩 represents 8 calculators

a Work out how many calculators were sold in October. **(1 mark)**
Use the information in parts b and c to help fill in the pictogram.
b The sales in January were the same as in November. **(1 mark)**
c The range for the sales is 16 calculators. The sales figure in December was the lowest. **(1 mark)**
d What is the mode? **(1 mark)**

2 **R** Here is a list of numbers.
13, 17, 12, 19
a What is the range of the numbers?
A number is added to the list.
The new range is 9.
b What number was added to the list? Show that there are two possible solutions to this problem.

3 **R** Four children stand in a row. The height of the shortest child is 1.2 m and the height of the tallest child is 1.5 m.

   a Which of the following could be the mean height of the children? Why?

   A 0.3 m **B** 1.2 m **C** 1.3 m **D** 1.5 m **E** 2.7 m

   b Assuming your answer to part **a** is correct, what could be the heights of the other two children?

4 The total weight of 5 pellets is 360 g. One pellet is removed and the mean weight of the remaining pellets is 75 g.

   a What is the weight of the pellet that was removed?

   b Was the weight of the pellet that was removed below or above the mean weight? Show how you know.

5 Dan wants to buy a new pair of skis.

   Shop A charges £480 but gives 15% discount for club members.

   Shop B charges £660, and is having a sale giving $\frac{1}{3}$ off.

   If Dan pays £25 to join the club, which shop offers the better deal?

6 Dean is comparing the data on the numbers of siblings (brothers and sisters) that students in two classes have.

   Both classes have 25 students.

   The tables show his results.

No. of siblings in class A	Frequency
0	1
1	9
2	8
3	4
4	3

No. of siblings in class B	Frequency
0	0
1	8
2	14
3	2
4	0
5	1

   Dean says, 'I can see from the tables that the average number of siblings in class A is 1 while in class B the average number of siblings is 2.'

   a Is he correct? Explain.

   Dean also says, 'Therefore the mean for class B is higher than that of class A.'

   b Is he correct? Explain.

   c What is the mean number of siblings in each class?

   d What is the median number of siblings in each class?

   e Write two statements to compare the data.

   > **Q6e hint** Sketching a dual bar chart may help.

7

These are the scores that Lizzie got in her last 9 practice papers.

36, 44, 58, 66, 34, 70, 78, 64, 80

What is her score on her 10th paper if

   a the range of her 10 scores is 51 **(1 mark)**

   b the median of her 10 scores is 62 **(2 marks)**

   c the mean of her 10 scores is 62? **(2 marks)**

8 **R** Here is a set of data arranged in ascending order: $x, x + 1, x + 1, 3x - 7, 4x + 5$

   a Work out the range, mode, median and mean for the set of data.

   Another number, bigger than $4x + 5$, is included in the data.

   b What is the new median?

9 **R** The table shows the numbers of letters received by post in a week by the families in one street.

Number of letters, x	Number of families
$0 \leqslant x < 4$	6
$4 \leqslant x < 8$	10
$8 \leqslant x < 12$	10
$12 \leqslant x < 16$	9
$16 \leqslant x < 20$	5

   a Calculate an estimate for the mean number of letters that is received per family on the street.

   b One family was not recorded in the table. When they are included in the data, the new estimate for the mean is 9.9. How many letters did this family receive in that week?

   c How will this entry change the table above?

10 **R** The missing frequencies for the table below are 1, 4, 6, 8 and 12, but not necessarily in that order.

Age (years)	Frequency
$10 < x \leqslant 15$	
$15 < x \leqslant 20$	
$20 < x \leqslant 25$	
$25 < x \leqslant 30$	
$30 < x \leqslant 35$	

   Copy the table three times and complete the tables so that

   a $15 < x \leqslant 20$ is the mode

   b $25 < x \leqslant 30$ is the median

   c you get the highest possible estimate of the mean. Show that the estimated mean will be 26.7 years (1 d.p.).

8 PERIMETER, AREA AND VOLUME 1

8.1 Rectangles, parallelograms and triangles

1 Calculate the area of each parallelogram.

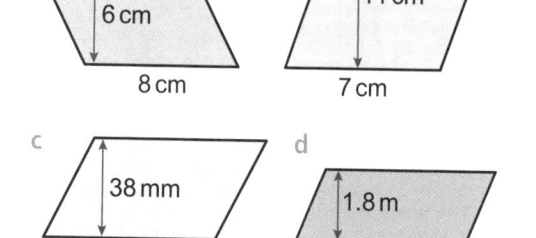

a 6 cm, 8 cm

b 11 cm, 7 cm

c 38 mm, 61 mm

d 1.8 m, 4.3 m

2 Calculate the perimeter and area of each shape. All lengths are in centimetres.

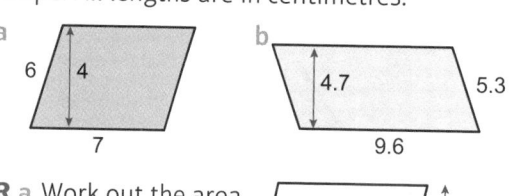

a 6, 4, 7

b 4.7, 5.3, 9.6

3 **R** a Work out the area of this parallelogram.

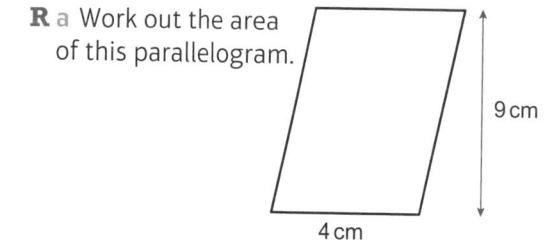

9 cm, 4 cm

b Trace the parallelogram. Join opposite vertices with a straight line to make two triangles.

c What fraction of the parallelogram is each triangle?

d Write down the area of one of the triangles.

4 Calculate the area of each triangle.

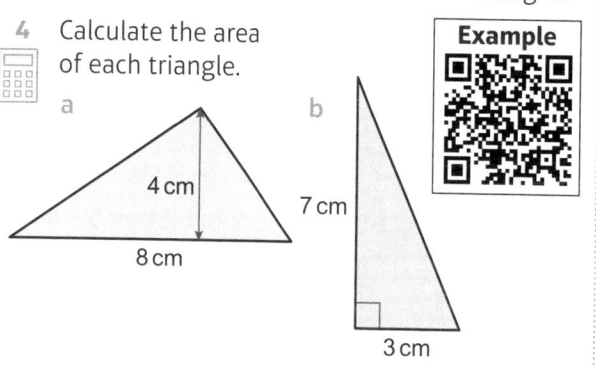

a 4 cm, 8 cm

b 7 cm, 3 cm

Example

c 2.6 cm, 6.7 cm

d 63 mm, 41 mm, 75 mm

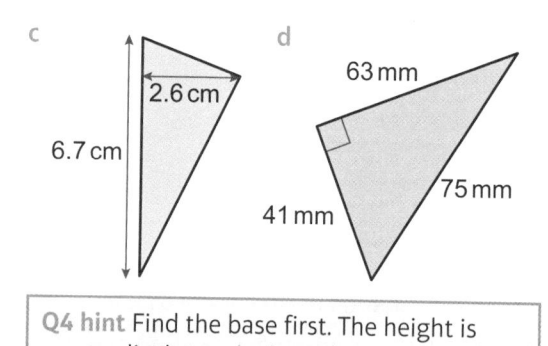

> **Q4 hint** Find the base first. The height is perpendicular to the base.

5 Calculate the perimeter and area of each shape. All lengths are in centimetres.

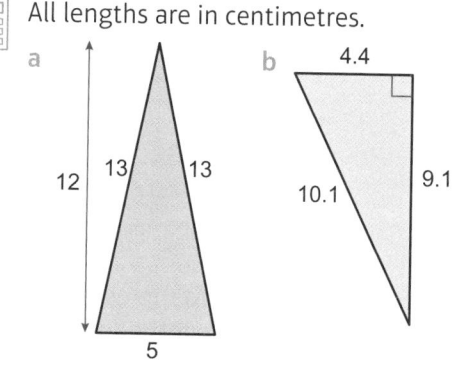

a 12, 13, 13, 5

b 4.4, 10.1, 9.1

6 A triangle has base length 5.6 cm and height 9 cm.
Calculate its area.

7 **P** Sam makes place mats from triangles cut from a wooden board.
Each place mat has a border of veneer around the whole triangle.

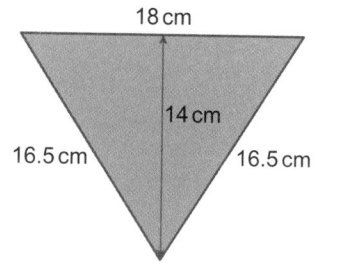

18 cm, 14 cm, 16.5 cm, 16.5 cm

He makes 15 of these triangles.
Work out
a the area of wood he needs
b the length of veneer he needs.

8 **P** Sketch and label three triangles that have area 12 cm².

9

Exam-style question

These shapes are drawn on centimetre squared paper.
Calculate the area of each one. **(6 marks)**

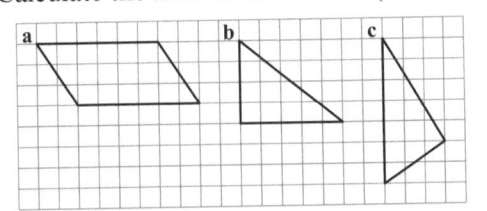

10 P Each shape has an area of 30 cm².
Calculate the lengths marked by letters.

a   7.5 cm   b

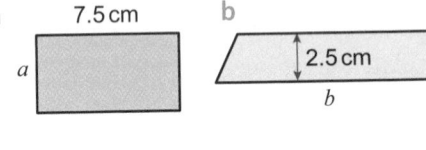

c

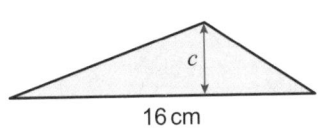

16 cm

11 A rectangular room measures 4.5 m by 5.1 m, and is 2.4 m high.
  a Sketch four rectangles to represent the four walls. Label their lengths and widths.
  b Calculate the total area of the walls.
  c A tin of paint costs £8.99 and covers 5 m². Estimate the cost of the paint for the walls.

8.2 Trapezia and changing units

1 Calculate the areas of these trapezia.
All lengths are in centimetres.
Round answers to 1 decimal place where necessary.

Example

a   7

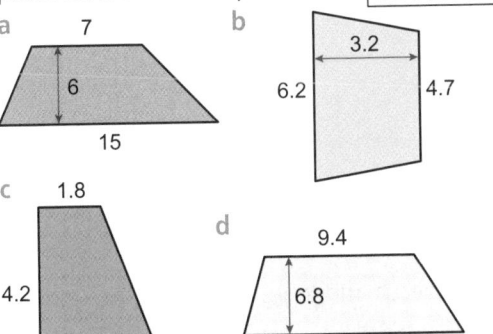

6

15

b

3.2

6.2   4.7

c   1.8

4.2

3.4

d

9.4

6.8

13.9

2 Calculate the area and perimeter of this isosceles trapezium.

6 cm

5.4 cm   5 cm

10 cm

> **Q2 hint** An isosceles trapezium has one line of symmetry. Its two sloping sides are equal.

3 P This trapezium has area 36 cm².

7 cm

h

11 cm

  a Substitute the values for A, a and b into the formula $A = \frac{1}{2}(a + b)h$
  b Simplify your answer to part **a**.
  c Solve your equation from part **b** to find h. Give the units with your answer.

4 P This trapezium has area 21 m².

6 m

h

8 m

Work out its height.

5 Convert
  a 1850 mm² to cm²   b 29.2 cm² to mm²

6 a Calculate the area of this square in cm².
    b Convert your answer to mm².

3.2 cm

7 R a Use these diagrams to help you work out the number of cm² in 4 m².

2 m

2 m

200 cm

200 cm

  b Copy and complete the double number line.

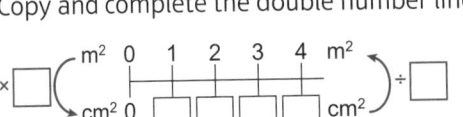

8 Convert

　a 5.8 m² to cm²　b 12 500 cm² to m²

9 a Choose four decimal measures between 3 m² and 4 m².

　Convert each measure to cm².

　b Choose four measures between 18 000 cm² and 20 000 cm².

　Convert each measure to m².

10 The diagram shows a paving stone.

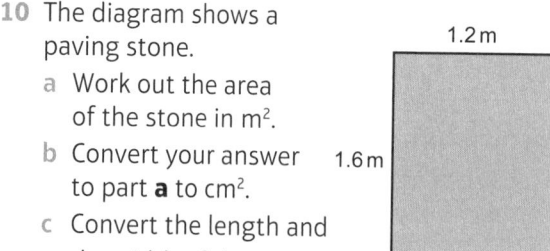

　a Work out the area of the stone in m².

　b Convert your answer to part **a** to cm².

　c Convert the length and the width of the paving stone to cm.

　d Use these measurements to work out the area of the paving stone in cm².

11 **P** Work out these areas in cm².

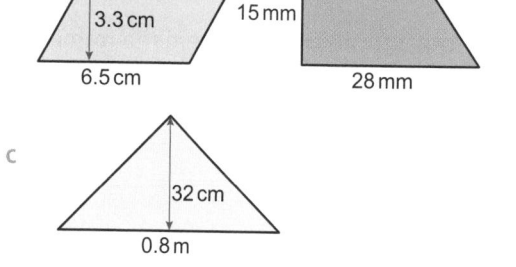

　a

　b　18 mm

　15 mm

　28 mm

　3.3 cm

　6.5 cm

　c

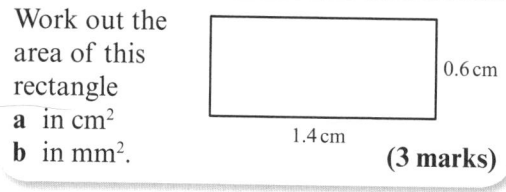

　32 cm

　0.8 m

12

Exam-style question

Work out the area of this rectangle

　0.6 cm

　1.4 cm

　a in cm²

　b in mm². **(3 marks)**

13 **P** Aisha is doing a wildlife survey.

In field A she finds 40 species in an area of 1.4 m².

In field B she finds 25 species in an area of 8000 cm².

Which field has more species per square metre?

Show working to explain.

8.3 Area of compound shapes

1 Convert these areas to the units given.

　a 5 ha to m²　　　b 7.5 ha to m²

　c 80 000 m² to ha　d 125 000 m² to ha

2 a Use these diagrams to help you work out the number of m² in 4 km².

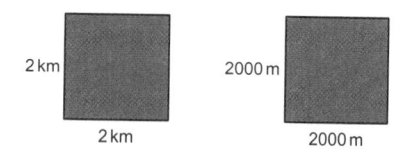

　2 km　　2 km

　2000 m　　2000 m

　b Copy and complete the double number line.

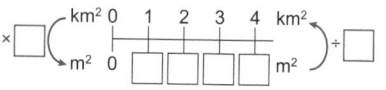

3 **R** Iona and Oronsay are islands of the Inner Hebrides.

The area of Iona is 8.77 km².

The area of Oronsay is 519 ha.

　a Which is larger, Iona or Oronsay?

　b What is the difference in area?

4 The diagram shows a shape split into two rectangles, A and B.

Example

　6 cm

　A

　8 cm

　B　5 cm

　10 cm

Work out

　a the areas of A and B　b the total area

　c the total perimeter.

5 Calculate the area and perimeter of these compound shapes.

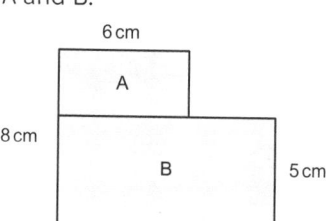

　a　5 cm

　9 cm

　3 cm

　7 cm

　b　5.8 cm　1 cm

　4.3 cm　3.6 cm

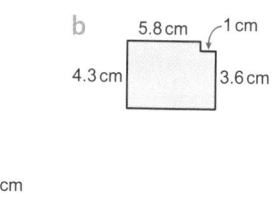

　c

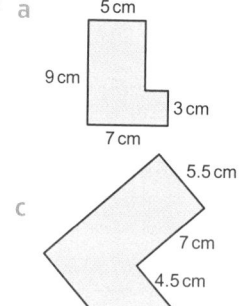

　5.5 cm

　7 cm

　4.5 cm

　5 cm

6 Here is a compound shape. There are two ways of working out its area.

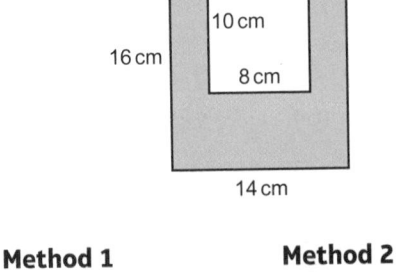

Method 1
Work out the area of the whole rectangle and subtract area A.

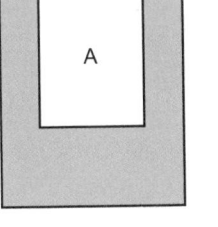

Method 2
Work out the areas of B, C and D and add them together.

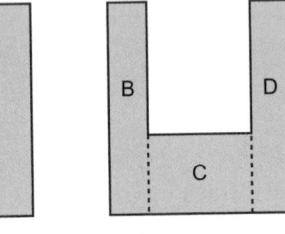

a Work out the area using both methods.
b Which do you prefer? Explain why.

7 Calculate the area and perimeter of each shape.

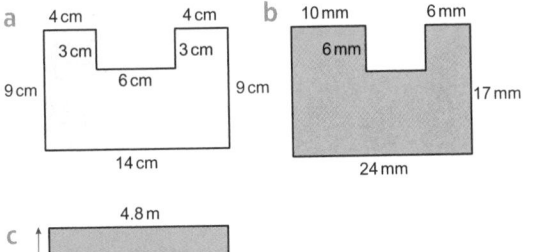

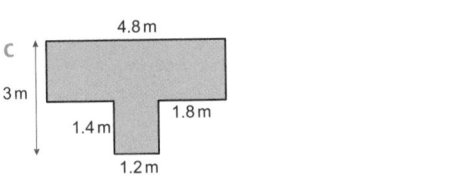

8 Work out the area of each shape.

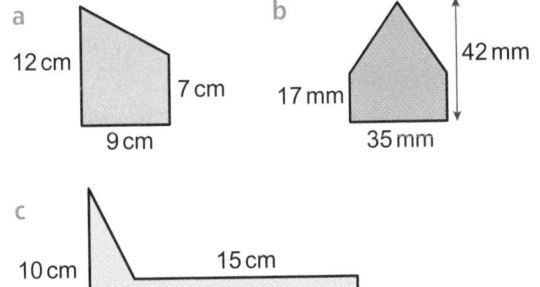

9 a Sketch this isosceles trapezium.

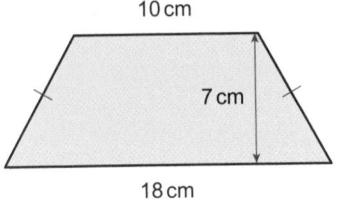

b Divide it into two triangles and a rectangle.
c Work out the area of each part, then add to find the total area.
d How could you work out the area of a trapezium if you forget the area formula?

10 **P** Rose makes this photo frame by cutting a rectangle 20 cm by 15 cm out of a larger rectangle of card.

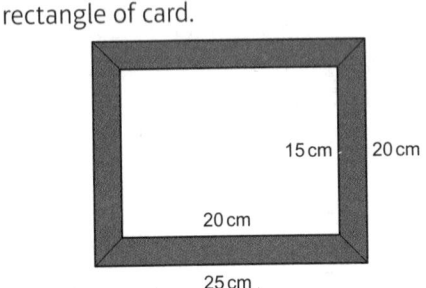

Work out the area of card in the frame.

> **Q10 hint** Work out the area of the larger rectangle then subtract the area of the cut out rectangle.

11 ⎰ **Exam-style question**

This window consists of a wooden frame with two panes of glass.

Work out the area of wood used.

(3 marks)

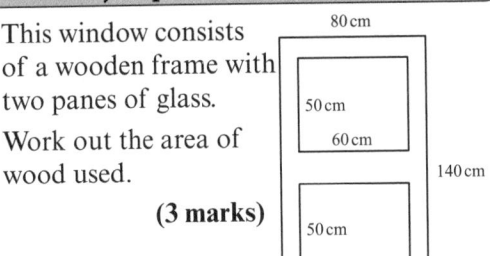

8.4 Surface area of 3D solids

1 Work out the surface area of this cuboid.

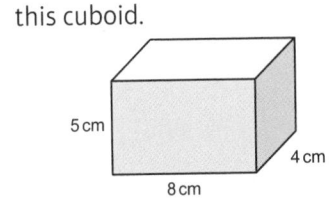

Example

2 R a Look at this diagram of a cuboid.

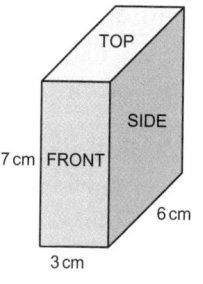

i What is the area of the top of the cuboid?
ii Which other face of the cuboid is identical to this one?

b i What is the area of the front of this cuboid?
ii Which other face of the cuboid is identical to this one?

c i What is the area of the side of this cuboid?
ii Which other face of the cuboid is identical to this one?

d Copy and complete this calculation to work out the total surface area of the cuboid.
$2 \times \square + 2 \times \square + 2 \times \square = \square\,cm^2$

3 Calculate the surface area of each cuboid.

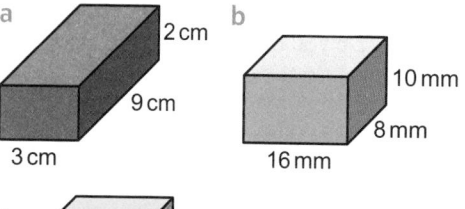

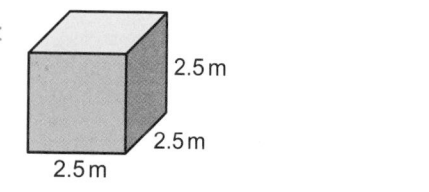

4 a Sketch a net of this triangular prism. Label the lengths.

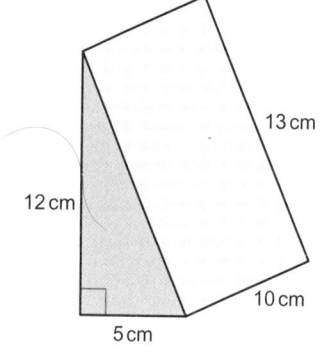

b Work out
i the area of each face
ii the total surface area.

5 Calculate the surface area of each 3D solid.

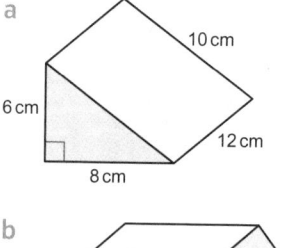

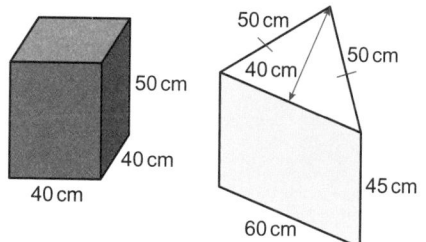

6 P The stand for a model is in the shape of a cuboid.
It is 65 mm long, 52 mm wide and 25 mm tall.
Ellie varnishes all the faces except the bottom.

a Work out the total area she varnishes, in square centimetres.

b One tin of varnish covers 120 cm². How many tins of varnish does Ellie need to varnish 5 model stands?

7 These stools are covered with fabric.

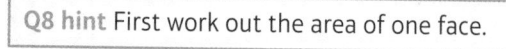

a Work out how many square metres of fabric you will need to cover them both.

The fabric costs £7.49 per square metre.

b Estimate the cost of the fabric needed to cover them both.

8 P / R This cube has surface area 726 cm².
What is the length of one side of the cube?

Q8 hint First work out the area of one face.

9

Exam-style question

Storage boxes are 60 cm wooden cubes, painted on all faces.

a Work out the area to be painted on one box.

b One tin of paint covers an area of 25 m². How many whole boxes can be painted with one tin of paint? **(4 marks)**

8.5 Volume of prisms

1. This cuboid is made of centimetre cubes.

   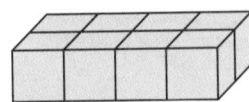

   Each cube has volume 1 cm³.
   Work out the volume of the cuboid.

2. Here are some more cuboids made of 1 cm³ cubes.
   Work out their volumes.

   a   b   c

   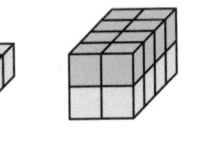

3. Work out the volume of each cuboid.
   Give the correct units with your answer.

   a   b

   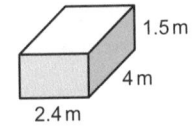

   3 cm, 4 cm, 7 cm   1.5 m, 4 m, 2.4 m

   c   d

   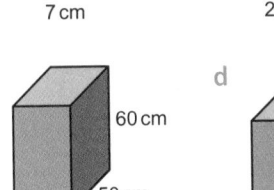

   60 cm, 50 cm, 35 cm   12 mm, 12 mm, 12 mm

4. This prism is made from centimetre cubes.
   Work out the volume of the prism.

   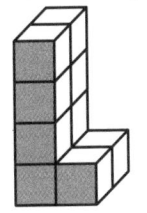

5. Calculate the volume of each prism.

   Example

   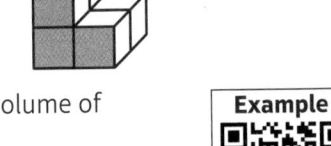

   a

   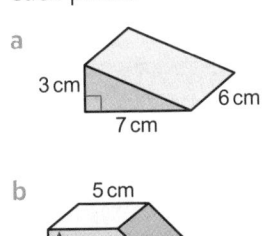

   3 cm, 6 cm, 7 cm

   b   c

   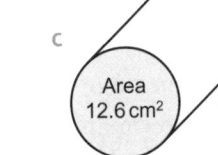

   5 cm, 4 cm, 9 cm, 3 cm   Area 12.6 cm², 6 cm

6. **P / R** The cross-section of a prism is a pentagon with area 8.4 cm².
   The prism is 5.3 cm long.
   Work out the volume of the prism.

7.

   The diagram shows a water tank in the shape of a cuboid.

   The tank is 1.8 m tall, 80 cm long and 60 cm wide.

   The depth of water in the tank is 1.4 m.

   1.8 m, 80 cm, 60 cm

   a How many cubic metres of water are in the tank? Give your answer to 2 decimal places. **(3 marks)**

   b Water costs £5.59 per cubic metre. How much will it cost to fill the rest of the tank? **(3 marks)**

   Q7 hint Mark the depth of the water on the diagram. Before you start your working out, check what units you need to use for your answer. Do you need to change any units before you do the calculation?

8. **P** A cube has volume 729 cm³.
   How long is one side of the cube?

9. **P** Sketch and label the dimensions of three cuboids with volume 36 cm³.

8.6 More volume and surface area

1. Convert
   a 17 cm³ to mm³   b 25 000 mm³ to cm³
   c 6.1 cm³ to mm³   d 12 870 mm³ to cm³

   Q1 hint
   ×1000
   cm³ ⟷ mm³
   ÷1000

2. a Work out the volume of this cuboid in cm³, then convert it to mm³.

   3.2 cm, 2.8 cm, 4.5 cm

   b Convert the measurements of the cuboid to mm, then work out the volume in mm³.

3 Convert
  a 7.2 m³ into cm³
  b 1.65 m³ into cm³
  c 2 500 000 cm³ into m³
  d 785 000 cm³ into m³

4 Convert
  a 0.8 litres into cm³   b 6600 cm³ into litres
  c 12 m³ into cm³   d 4 m³ into litres
  e 1.9 m³ into litres   f 8000 litres into m³
  g 5.3 litres into cm³
  h Copy and complete:
    1 m³ = ☐ litres

5 This measuring cylinder can hold 100 ml.
  What is the capacity of this measuring cylinder in cm³?

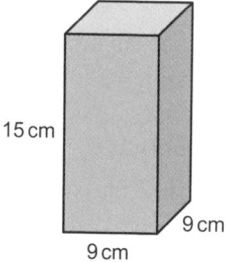

6 A raised bed for growing vegetables is in the shape of a cuboid.

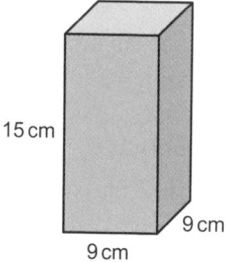

30 cm   70 cm
120 cm

Phoebe wants to fill it with compost.
How many litres of compost does she need?

7 Work out the capacity of this carton
  a in ml
  b in litres.

15 cm
9 cm
9 cm

8 **P** Here is the design for a garden pond.

2.5 m
1.8 m   0.7 m
2.6 m

a How much water will the pond contain?
  Give your answer in litres.
The bottom and sides of the pond are covered with a plastic lining.
b How many square metres of lining are needed?

9 Alex pours melted wax into moulds like this to make candles.

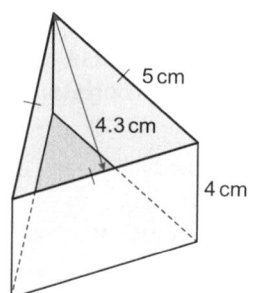

5 cm
4.3 cm
4 cm

a What volume of wax does he need for each candle?
Alex melts this cube of wax.

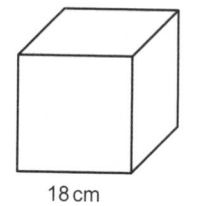

18 cm

b How many candles can he make with this wax?

10 Work out the missing measurements in these cuboids.

Example

a

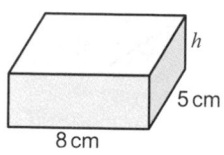

h
5 cm
8 cm
Volume = 120 cm³

b

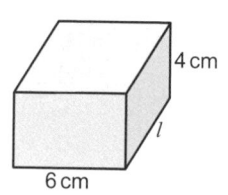

4 cm
l
6 cm
Volume = 336 cm³

11 **Exam-style question**

This cuboid has volume 480 cm³.

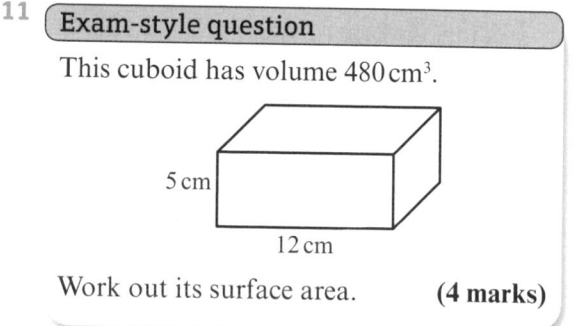

5 cm
12 cm

Work out its surface area.   **(4 marks)**

8 Problem-solving

Solve problems using these strategies where appropriate:

Example

- **Use pictures**
- **Use smaller numbers**
- **Use bar models**
- **Use x for the unknown**
- **Use a flow diagram.**

1 **R** A square is cut in half to form two identical rectangles. The rectangles have one side that is 6 cm long.
  a What could the length of the other side be? (There is more than one possible answer.)
  b What could the area of the square be?

2 Nicola buys a new pair of running shoes and a T-shirt. She spent £86 on the shoes. This is £23 more than three times the cost of the T-shirt.
How much did the T-shirt cost?

3 **R** This trapezium is made up of a square and a triangle of the same area.

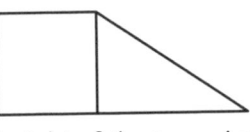

What is the height of the trapezium if its area is 242 cm²?

4 A small field has a width that is 4.2 m shorter than its length. The perimeter is 75.6 m. What are the dimensions of the field?

5 **Exam-style question**

Jenny is redecorating her bathroom, which is 2.8 m wide by 3.5 m long and 3 m high.

She tiles one of the smaller walls and paints the other 3 walls.

The tiles are 35 cm by 70 cm and are sold in boxes of 10.

Each box costs £15.

Paint is sold in tins which cost £8.99 and cover 4.5 square metres.

Which will cost her more, the tiles or the paint?

How much more?

(6 marks)

6 In the shape below all lengths are in cm.

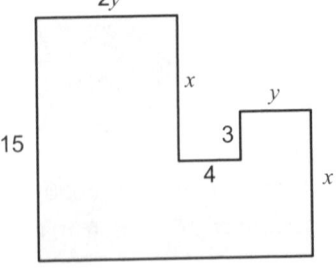

  a Write the down the perimeter of this shape in terms of x and y.
  b Show that $x = 9$
  c If $x = 2y$, what is the area of the shape?

7 At an athletics event, $\frac{3}{5}$ of the crowd are adults. 75% of the remaining crowd are girls. There are 4500 girls in the crowd. How many people in total are in the crowd at the athletics event?

> **Q7 hint** Draw a bar to represent all the crowd. Split the bar into 10 equal sections and work out what 1 section represents.

8 **R** Casey has 60 jewellery boxes of dimensions 6 cm by 4.5 cm by 2 cm. She packs all of them neatly into a crate. There is no space left in the crate.
  a What is the capacity of the crate?
  b Suggest the dimensions of a suitable crate.

9 **R** A butter manufacturer wants to sell butter in 48 cm³ portions. A manager chooses to package it in 2 cm by 4 cm by 6 cm slabs. Suggest different dimensions so that less material is needed to wrap each slab of butter.

10 **R** A cylindrical tank of height 0.8 m and cross-sectional area 0.12 m² is $\frac{3}{4}$ full of oil.
Is there enough oil in the tank to fill 85 half-litre bottles of oil?
How can you tell?

11 The cross-sectional area of shape A is 28 cm². It has a volume that is $1\frac{1}{2}$ times that of shape B. What is the length x of shape A?

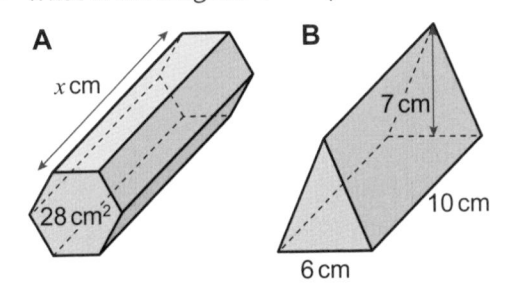

9 GRAPHS

9.1 Coordinates

1 R a Angie uses this rule to generate coordinates.

The x-coordinate is always 3, no matter what the y-coordinate is.

Which of these coordinate pairs satisfy Angie's rule?

$(3, 2)$, $(-3, 5)$, $(0, 3)$, $(3, 3)$, $(4, 3)$, $(-1, 3)$, $(3, -2)$

b Draw a coordinate grid from −5 to +5 on both axes.

Plot the points from part **a** that satisfy Angie's rule.

What do you notice about the points you have plotted?

c Henry uses this rule to generate coordinates.

The x-coordinate is always 5, for any y-coordinate.

Henry generates the coordinates $(5, 2)$, $(5, -1)$, $(5, 0)$ and $(5, 4)$.

Where do you expect these points to be on the grid?

d Plot the points on the same grid. Were you correct?

2 a Write down the integer coordinates of five points on line P.

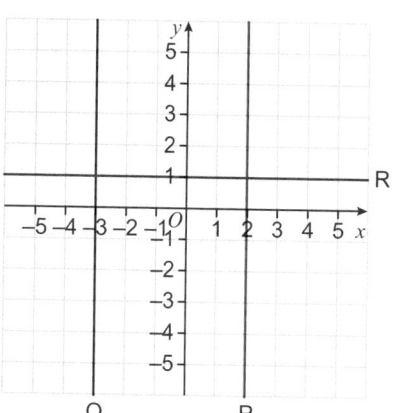

b What do you notice about the coordinates on line P?

c Copy and complete.

  i The equation of line P is $x = \ldots$

  ii The equation of line Q is $x = \ldots$

  iii The equation of line R is $y = \ldots$

3 Write the equations of the lines labelled A, B, C and D.

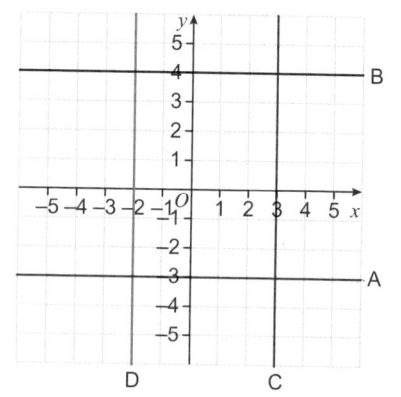

4 R Draw a coordinate grid from −5 to +5 on both axes.

Draw and label these graphs.

a $y = 2$ **b** $x = -1$ **c** $y = -3$

d What is the equation of the y-axis?

> **Q4d hint** What is the x-coordinate of every point on the y-axis?

5 R a Write down the integer coordinates of the points on this line.

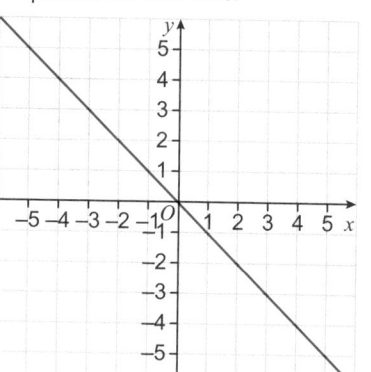

b What do you notice about the coordinates?

c Complete the missing coordinates of these points on the line.

$(6, \square)$, $(\square, -8)$

d The equation of the line is $y = \ldots$

6 These points are on the line $y = x$.

$(-3, -3)$, $(-2, -2)$, $(0, 0)$, $(1, 1)$, $(5, 5)$

a Draw a coordinate grid from −5 to +5 on both axes.

b Plot the points and join them with a straight line.

c Write the coordinates of three more points on the line $y = x$.

7 Find the midpoint of each of these line segments. Make a table to help you.

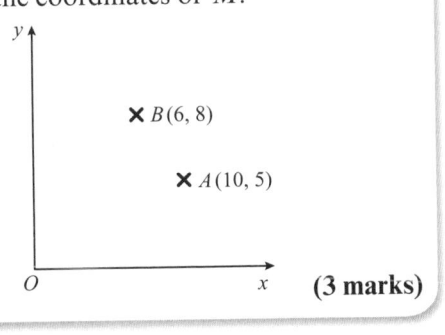

Line segment	Start point	End point	Midpoint
AB	(2, 3)	(2, 5)	(2, 4)
CD			
EF			
GH			
JK			
LM			
NP			

8

> **Exam-style question**
>
> The point A has coordinates (10, 5).
>
> The point B has coordinates (6, 8).
>
> M is the midpoint of AB.
>
> Find the coordinates of M.
>
> **Example**
>
> [QR code]
>
> [graph with points ✗ B(6, 8) and ✗ A(10, 5), axes y and x, origin O]
>
> **(3 marks)**

9 Work out the midpoints of the line segments with these start and end points.

a (1, 8) and (7, 4)

b (1, 3) and (6, 8)

c (−2, 5) and (4, 1)

d (−3, −7) and (0, −2)

9.2 Linear graphs

1 This function machine generates coordinates.

input function output coordinates

$x \rightarrow$ [ ] $\rightarrow y \rightarrow (x, y)$

$-2 \rightarrow$ [−3] $\rightarrow$ ☐ $\rightarrow$ (−2, ☐)

$0 \rightarrow$ $\rightarrow$ ☐ $\rightarrow$ (0, ☐)

$4 \rightarrow$ $\rightarrow$ ☐ $\rightarrow$ (4, ☐)

a Work out the missing coordinates.

b Draw a coordinate grid from −5 to +5 on both axes.

c Plot the coordinates from part **a** on the grid.

d Join the points and extend the line to the edges of the grid.

e Write the coordinates of two more points that lie on the line.

f Write a rule using algebra for this function machine.

g Label your graph with this rule.

> **Q1f hint** Your rule should start $y = \ldots$
> What do you do to the x-value to get the y-value?

2 Repeat **Q1** for this two-step function machine. Use a coordinate grid from −2 to 8 on both axes.

input functions output coordinates

$x \rightarrow$ $\rightarrow y \rightarrow (x, y)$

$-1 \rightarrow$ [×2] [+3] $\rightarrow$ ☐ $\rightarrow$ (−1, ☐)

$0 \rightarrow$ $\rightarrow$ ☐ $\rightarrow$ (0, ☐)

$2 \rightarrow$ $\rightarrow$ ☐ $\rightarrow$ (2, ☐)

3 a Copy and complete the tables of values for these straight-line graphs.

Example

[QR code]

x			−3	−2	−1	0	1	2	3
$y = x - 1$			−4			−1			

x			−3	−2	−1	0	1	2	3
$y = 2x + 2$			−4			2			

b Draw a coordinate grid with −3 to +3 on the x-axis and −8 to +8 on the y-axis. Draw and label the graphs of $y = x - 1$ and $y = 2x + 2$, using your tables of values from part **a**.

4. Draw and label these straight-line graphs for $x = -3$ to $+3$.
   Use a suitable coordinate grid.
   a $y = 2x - 1$ b $y = 3x - 2$
   c $y = 4x - 5$ d $y = 0.5x - 3$

5. a Copy and complete these tables of values for straight-line graphs.

   i

x	-3	-2	-1	0	1	2	3
$y = -x - 2$	1			-2			

   ii

x	-3	-2	-1	0	1	2	3
$y = -2x + 1$	7			1			-5

   iii

x	-3	-2	-1	0	1	2	3
$y = -3x + 2$	11			2			-7

   b Draw a coordinate grid with -3 to +3 on the x-axis and -11 to +11 on the y-axis. Draw and label the graphs of $y = -x - 2$, $y = -2x + 1$ and $y = -3x + 2$, using your tables of values from part **a**.

6. Draw and label the graphs of these straight lines on the same grid, for values of x from -4 to +4.
   a $y = -2x + 2$ b $y = -x - 1$
   c $y = -3x + 5$ d $y = -4x + 7$
   e Would you expect these graphs to slope upwards or downwards from left to right?
   i $y = 2x - 8$ ii $y = -3x + 2$

7. **Exam-style question**

   a Complete the table of values for $y = 3x - 1$.

x	-2	-1	0	1	2	3	4
y							

   b On the grid, draw the graph of $y = 3x - 1$.

   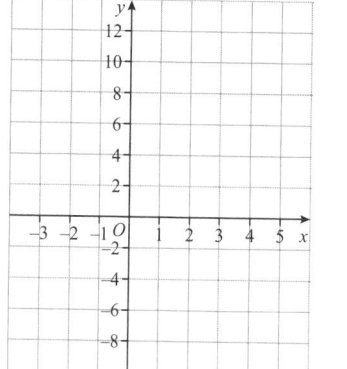

   (4 marks)

8. The diagram shows the graph of $y = 3 - \frac{1}{2}x$

   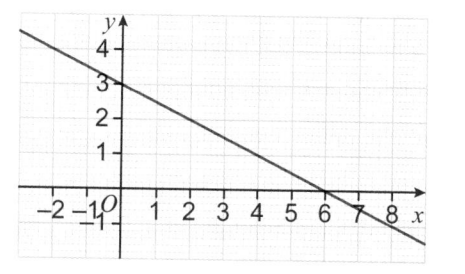

   a Use the graph to find an estimate for the value of y when
   i $x = 2.5$ ii $x = 6.8$ iii $x = -1.2$
   b Use the graph to find an estimate for the value of x when
   i $y = 2.1$ ii $y = 0.4$ iii $y = -0.8$
   c Use the graph to find approximate solutions to these equations.

   i $3.2 = 3 - \frac{1}{2}x$

   ii $1.7 = 3 - \frac{1}{2}x$

   iii $2.4 = 3 - \frac{1}{2}x$

   Q8a i hint Find the point on the line where $x = 2.5$ and read across to the y-axis to find the value of y.

9.3 Gradient

1. Look at the graph.

   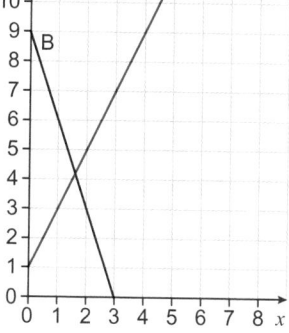

   a How many squares does line A go up for every
   i 1 square across
   ii 2 squares across
   iii 3 squares across?
   b How many squares does line B go down for every
   i 1 square across
   ii 2 squares across
   iii 3 squares across?

2 Here are the graphs of $y = 3x - 2$, $y = 3x + 1$ and $y = -3x + 4$

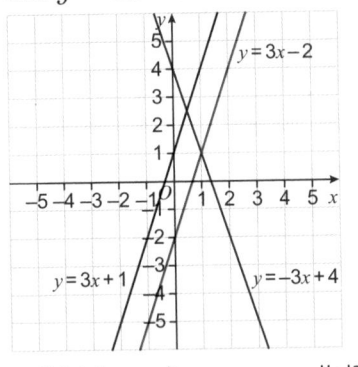

a Which two lines are parallel?

b What are the gradients of the parallel lines?

3 Work out the gradient of each line segment by calculating $\dfrac{\text{total distance up}}{\text{total distance across}}$

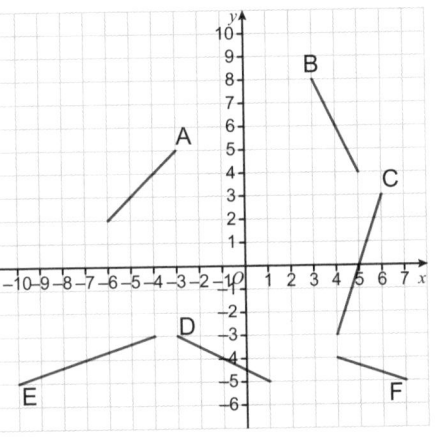

4 On squared paper draw line segments with these gradients.

a 2   b −1   c −2   d $\frac{1}{4}$ e $-\frac{1}{2}$

5 **Exam-style question**

Find the gradient of the straight line drawn on this grid.

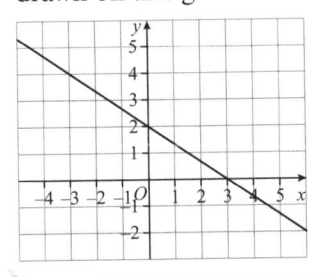

(2 marks)

6 These sketches show ramps on an assault course. Work out the gradient of each ramp.

a
0.1 m
0.5 m

b
0.14 m
7 m

c
3.5 m
14 m

7 a Copy and complete these tables of values.

i
x	−3	−2	−1	0	1	2	3
$y = x - 2$	−5			−2			

ii
x	−3	−2	−1	0	1	2	3
$y = 2x + 1$	−5			1			

b Draw the graphs of

i $y = x - 2$ ii $y = 2x + 1$

c Work out the gradient of each line.

8 **R** Choose from these equations

$y = 2x + 1$ $y = -3x - 2$ $y = -2x + 1$
$y = 4x - 3$ $y = x + 5$ $y = 2x - 3$

a the line with the steepest gradient

b a pair of parallel lines

c a line that slopes down from left to right.

Example

9 Find the gradient of the line joining the points (3, 1) and (7, 3).

9.4 $y = mx + c$

1 The diagram shows five straight-line graphs and their equations.

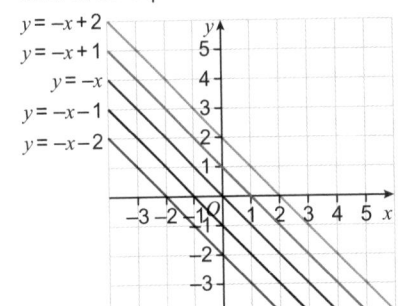

a What do you notice about the equation of the line and the y-intercept?

b Where would you expect $y = -x + 4$ to cross the y-axis?

c What is the gradient of each line?

2 Work out the equations of lines A, B, C, D and E.

Example

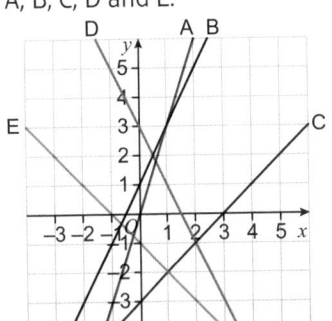

3 Write down the equation of a straight line that is parallel to $y = 3x - 4$

4 Which of these lines pass through
a (1, 1)
b (0, −2)?

$y = x$	$y = x + 2$	$y = 2x$
$y = 2x - 2$	$y = x - 2$	$y = 2x - 1$

5 In each of these you are given one point that lies on a line and the gradient of the line. Work out the equation of the line.
a (1, 2) and $m = 2$
b (3, 5) and $m = 3$
c (2, −1) and $m = 1$
d (−3, 4) and $m = \frac{1}{2}$
e (−3, −5) and $m = 5$

6 In each of these you are given two points that lie on a line.
Work out the equation of the line.
a (1, 3) and (3, 7)
b (2, 6) and (6, 8)
c (1, 1) and (3, −3)
d (−2, 3) and (−5, 6)

7 Draw these graphs from their equations.
a $y = x - 5$
b $y = -4x + 1$
c $y = \frac{1}{2}x + 2$
d $y = -2x + 1$

8 Draw a coordinate grid from 0 to 8 on both axes. Plot these graphs.
a $x + y = 3$
b $x + y = 5$
c $x + y = 2$
What do you notice?

9 Draw a coordinate grid from −8 to +8 on both axes. Plot these graphs.
a $2x + 3y = 6$
b $y - x = 8$
c $2x + 4y = 4$
d $3x - y = 9$

9.5 Real-life graphs

1 **P** a An electricity company charges £0.10 per kWh unit. Copy and complete the table below to show the costs of using electricity.

Number of units	0	100	200	300	400	500	600
Cost (£)	0	10					

b Draw a graph of units used against cost.
c How much would you pay if you used 150 units?
d Mr Bridge buys his electricity in advance. If he buys £20 worth and then uses 60 kWh units, how many units does he have left?

2 A mobile phone company charges a flat rate of £10 per month plus £5 for every 20 minutes of calls.
a Copy and complete the table.

Minutes of calls	0	20	40	60	80	100
Cost (£)	10					35

b Draw a graph of number of minutes of calls against cost.
c Anna's phone bill is £27. Use your graph to find the total length of calls she made.

3 **R** The graph shows the cost of travelling in a taxi.
a What is the cost of
  i a 50-mile journey
  ii a 10-mile journey?
b What is the gradient of the graph?

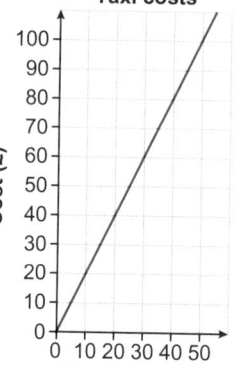

4 **R** The graphs show the costs of tiles at three different DIY shops.
Store A Tiles £4 each
Store B Tiles £8 each
Store C Tiles £6 each
a Match each line on the graph with the correct store.
b Copy and complete this statement.
The steeper the slope the ……….. the unit price.

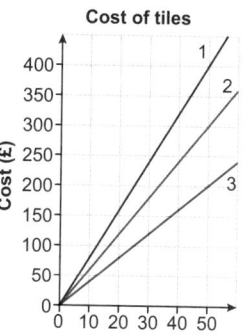

5 **R** A plumber charges a £50 callout fee and £25 per hour.
a Copy and complete the table of values.

Hours worked	0	1	2	3	4	5
Total cost (£)	50					175

b Draw a graph of hours worked against total cost.
c Work out the equation of the line.
d What does the gradient represent?
e What does the y-intercept represent?

6 The graph shows the costs of two different car hire companies.

Car hire costs

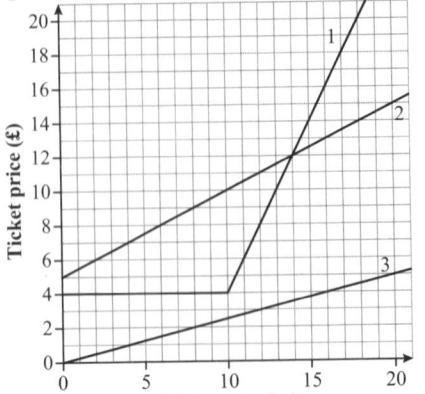

A2B Cars charge £50 flat rate and £0.50 per mile.
121 Cars charge £0.70 per mile.

a Match each line on the graph with the correct car hire company.

b Chaitaly is going on a 300-mile journey. Explain which car hire company she should use.

7 Obalesh has a movie package on his television. He pays £15 per month plus £3 for each film he watches. He watches about 6 films a month.
He sees a new offer for £25 per month plus £1 per film.

a Should he change to the new offer?

b How could you see at a glance when each payment option was better value?

8 The graph shows two different ways to pay monthly for a gym membership.

Gym membership fees

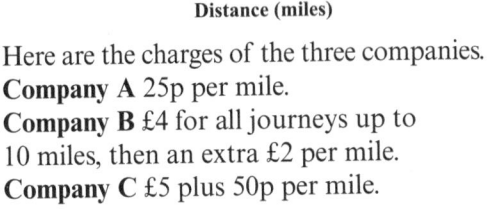

a Explain in words what each option means.

b Diccon expects to do approximately 15 classes a month.
Which option should he choose?

9 A car salesman is paid using this equation.

$$y = \frac{x}{100} + 1500$$

where y is his monthly wage (£) and x is the total value of all the cars he sells.

a Draw the graph for the equation.

b What is his monthly pay when no sales are made?

c He usually sells cars to the value of approximately £45 000 a month. He is offered a different job where there is a fixed salary of £3000 a month. Should he take it?

10

Exam-style question

The graph shows how three different bus companies work out the cost of their tickets.

Here are the charges of the three companies.
Company A 25p per mile.
Company B £4 for all journeys up to 10 miles, then an extra £2 per mile.
Company C £5 plus 50p per mile.

a Match each line on the graph to the letter of the company it represents.

Sanjit uses company B to travel 12 miles.

b Find the total cost of a ticket.

Maarta needs to travel 15 miles. She can only use company B or C.

c Explain which company would be cheaper for her to use.
You *must* give reasons for your answer.

(4 marks)

Exam hint

To get all the marks you need to refer to the numbers on the graph as you explain your reasons.

9.6 Distance–time graphs

1 **R** Aaban leaves for school at 8 am. He walks for 1 mile to his friend's house, which takes him 20 minutes.

He then meets his friend at his house, stops for 10 minutes for a cup of tea, and walks a further $\frac{1}{2}$ mile, which takes him 15 minutes.

Example

a Draw a distance–time graph for Aaban's journey.

b Work out the average speed in miles per hour for each part of his journey.

c Compare the average speeds to decide in which part of the journey he was walking more quickly.

> **Q1a hint** Draw a horizontal axis from 0 to 60 minutes and a vertical axis from 0 to 2 miles.

2 This distance–time graph shows Claudia's morning run.

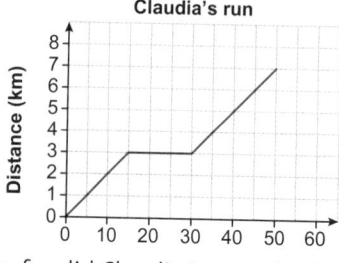

a How far did Claudia jog on the first part of the run?
b How long did she jog for before she stopped for a break?
c How long did she rest for?
d What was her average speed in km/h for the first part of the run?
e How long did the last part of the run take?
f How far did she jog on the last part of the run?
g What was her average speed for the last part of the run?

3 This graph shows Eiliyah's distance from home when walking her dog.

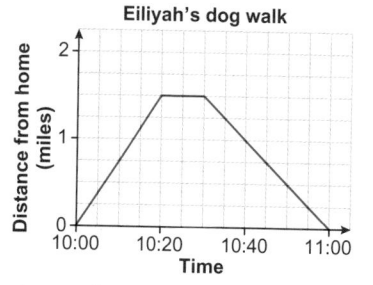

a Work out the average speed of her walk for the first part of the journey.
b How long did she stop for?
c Work out the average speed on her return journey.
d How far did she walk altogether?

4 Maaz leaves home at 9 am and drives 20 miles to work. It takes him $\frac{1}{2}$ hour to get there.
He stays at work for $\frac{1}{2}$ hour before getting into his car and driving 15 miles in 20 minutes to a call-out.
He stays there for $1\frac{3}{4}$ hours before returning home, arriving home at 1 pm.
a Draw a distance–time graph to show Maaz's journey.
b On which part of the journey was he travelling the fastest?

5 The graph shows two train journeys.
One train leaves Petersfield at 10 am and travels to Guildford. The other train leaves Guildford at 10 am and travels to Petersfield.

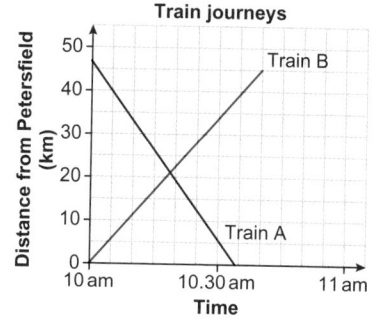

a Which line shows the Guildford to Petersfield train?
b What time does each train reach its destination?
c Which train journey is faster?

6 Angela and Wadi go scuba diving.
The graph shows their depths below sea level at different times.

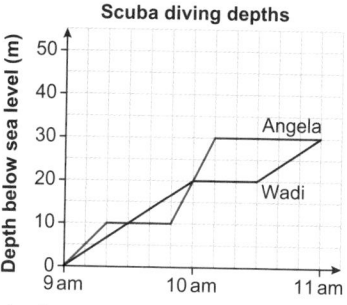

a What is the maximum depth that they dive to?
b How long does Wadi spend at 20 metres below sea level?
c Between what times is Wadi deeper than Angela?
d How long does it take Angela to reach her maximum depth?

7 Wabisa runs from her house to the gym. The travel graph shows her journey.

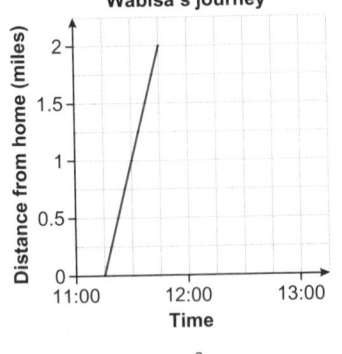

Wabisa's journey

She stays at the gym for $\frac{3}{4}$ hour. Then she jogs home, which takes her $\frac{3}{4}$ hour.

a Copy and complete the travel graph.
b What time does Wabisa leave home?
c How far is the gym from Wabisa's house?

8 Exam-style question

The graph shows part of a journey of a train travelling at constant speed.

Another train travels at a speed of 100 mph. Which train is faster?

You must explain your answer.   **(3 marks)**

9 Match each of these velocity–time graphs to one of the sentences.

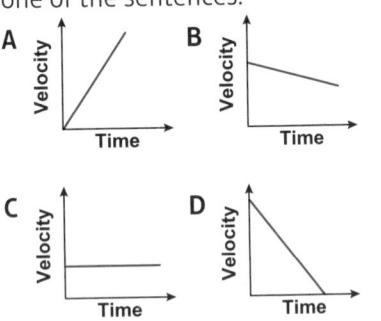

A   B

C   D

i Starts from rest and velocity steadily increases.
ii Travels at a steady velocity.
iii Velocity steadily decreases before stopping.
iv Steadily decreases in velocity.

Q9 hint **Velocity** means speed in a particular direction.

10 The graph shows the velocity of a car when it is braking.

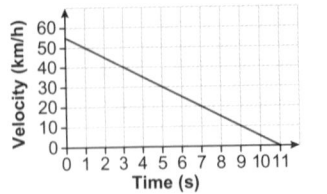

a What was the starting velocity of the car?
b What was the deceleration of the car?
c What is the equation of the line?

9.7 More real-life graphs

1 R The graph shows the depth of water for three containers A, B and C, filling at a constant rate.

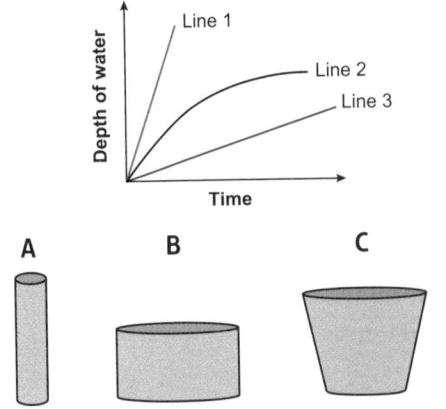

a Which container will fill fastest?
b Match each container to a line on the graph.

2 P a Two baths A and B are filled with water at a constant rate. Sketch a graph of depth of water against time for each of the baths.

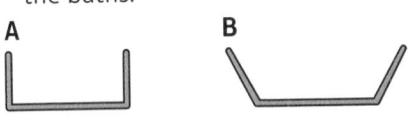

A   B

b Two different containers are filled at a constant rate. Match each container to its graph.

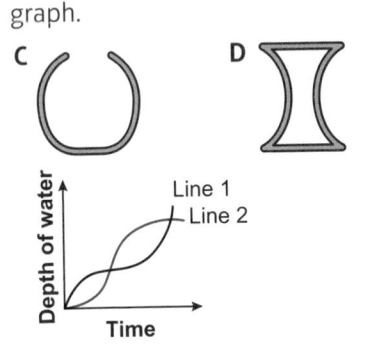

C   D

3 The graph shows the pressure on your eardrums (in atmospheres) at up to 20 m depth in water.

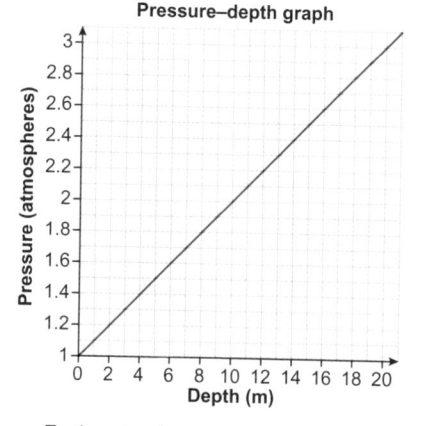

Pressure–depth graph

a Estimate the pressure at 10 m depth.

b Estimate the depth of water when the pressure is 1.2 atmospheres.

c Rabail says, 'The rate of increase of pressure with depth is always the same'. What evidence from the graph supports this?

d Work out the gradient of the line.

4

Exam-style question

Water fills a cuboid swimming pool at a constant rate. The graph shows how the depth of water in the pool varies over time.

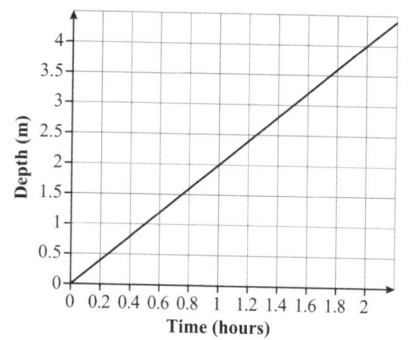

a Work out the gradient of the straight line.

b Write down a practical interpretation of the value you worked out in part **a**.   (3 marks)

5 **R** The scatter graph shows the height and weight of 6 children.

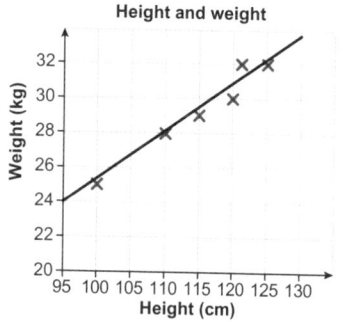

Height and weight

a Describe the correlation between the height and weight.

b Work out the equation of the line of best fit.

c Use your equation to predict the weight of a child who is 105 cm tall.

d Explain why you cannot estimate the weight of a child who is 140 cm tall.

Q5c hint Using the equation gives a more accurate value than reading from the graph.

6 The table shows the distance a car has travelled and its current value.

Distance travelled ('000 miles)	110	100	30	60	70	95	20
Value (£'000)	3	4	8	7	7	5	10

a Draw a scatter graph for the data.

b Draw a line of best fit.

c Find the equation of your line of best fit.

d A car has travelled 105 000 miles. Estimate the value of the car.

e A car is for sale at £7500. Estimate the distance it has travelled.

f A car has travelled 150 000 miles. Estimate the value of the car.

g How reliable is your estimate in part **f**?

9 Problem-solving

Solve problems using these strategies where appropriate:
- **Use pictures**
- **Use smaller numbers**
- **Use bar models**
- **Use x for the unknown**
- **Use a flow diagram.**

1 Cory uses the formula $y = 1.8x + 32$ to convert the temperature from Celsius to Fahrenheit for the past few days. The temperatures were 18°C, 16°C, 21°C and 25°C.

What are the temperatures Cory calculated in Fahrenheit, rounded to the nearest degree?

Q1 hint Can you use a diagram to show the function?

2 Charlie can buy a pay-as-you-go mobile plan or a flat-rate mobile plan. The graph shows two different options. Both options offer unlimited data, so Charlie is considering the number of minutes he will use on calls.

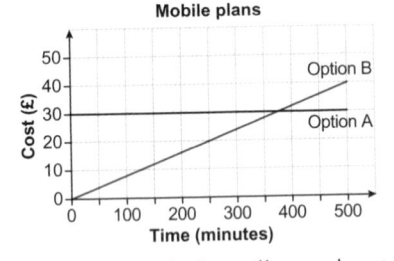

a Charlie thinks he will use about 300 minutes per month. If he is correct, which plan is cheaper for him to use?

b How much cheaper each month would this plan be?

3 Chen is running a bath. The water flows at a constant rate of 11 litres/minute.

a Draw a graph to show the water flow over 20 minutes.

Chen's bath has a capacity of 262 litres. She would like it to be half-filled before she gets in.

b Use the graph to help you find approximately how many minutes she will have to wait before she can get in.

4 Catriona has two spinners, A and B.

For spinner A, P(blue) = $\frac{4}{6}$

For spinner B, P(blue) = $\frac{5}{9}$

Which spinner is more likely to land on blue?

5 Delia walks from her house to her friend Claire's house. She then walks to the shop and returns home.

The graph shows Delia's journey.

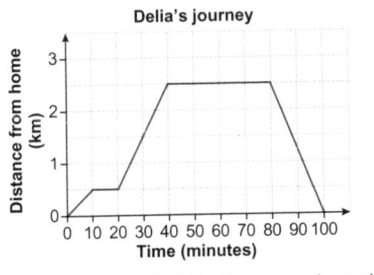

a How long did Delia spend at the shop?

b On which part of her journey did Delia walk the fastest?

c How far is the shop from Delia's house?

6 The diagram shows the plan of a playground. Lengths are in metres. The area of the playground is 300 m².

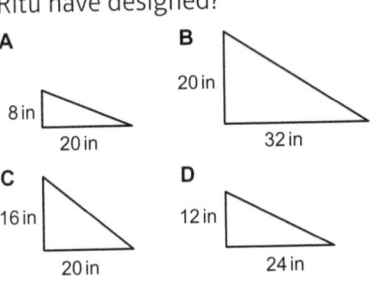

a Write an equation and solve it to find a.

b If the gate into the playground is 1.5 m wide, what length of fencing is needed to go round the rest of the playground?

7 **R** Ritu designs some mini skateboard ramps. The ramps she designs all have a gradient of 0.6 or more. Which of the ramps below could Ritu have designed?

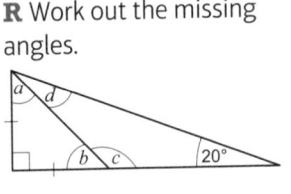

8

9 **R** Work out the missing angles.

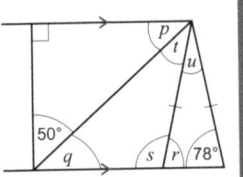

10 Kai wants to know if he will need to invest in a better advertising campaign next year. His company managed to reach their sales target of 900 books in 2014. In 2015, their target is 1500. The table shows the total sales figures from 2009 to 2014.

2009	2010	2011	2012	2013	2014
110	156	280	442	634	920

a Put this data into a graph. What do you notice? Is there a trend?

b Do you think Kai's company will make their target in 2015 or not? Explain.

10 TRANSFORMATIONS

10.1 Translation

1   Copy the diagram.

Example

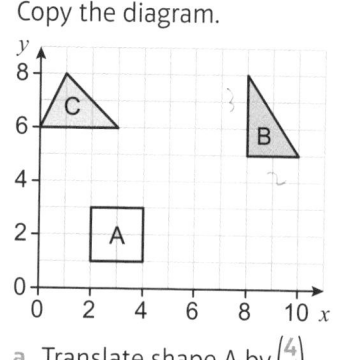

   a   Translate shape A by $\begin{pmatrix} 4 \\ 2 \end{pmatrix}$.

   b   Translate shape B by $\begin{pmatrix} -3 \\ 2 \end{pmatrix}$.

   c   Translate shape C by $\begin{pmatrix} 2 \\ -4 \end{pmatrix}$.

2   Copy the diagram.

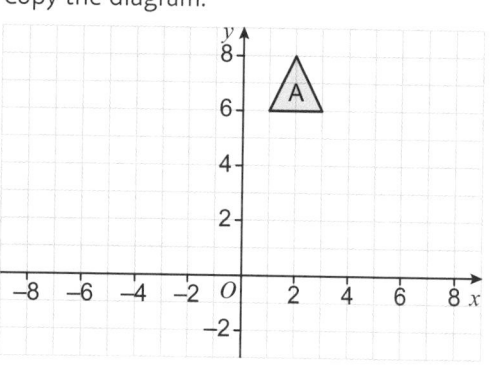

   Translate shape A by each of these column vectors. Label the images B to F.

   a $\begin{pmatrix} 2 \\ 1 \end{pmatrix}$ b $\begin{pmatrix} 3 \\ -2 \end{pmatrix}$ c $\begin{pmatrix} -6 \\ 0 \end{pmatrix}$ d $\begin{pmatrix} -4 \\ -2 \end{pmatrix}$ e $\begin{pmatrix} 0 \\ -4 \end{pmatrix}$

3   Write down the column vector that translates shape A to shape B.

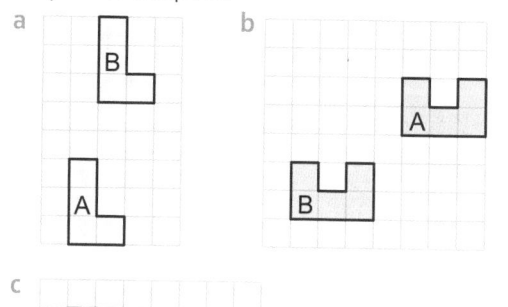

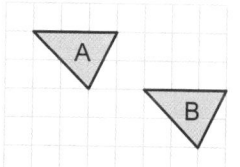

4   Look at this diagram.

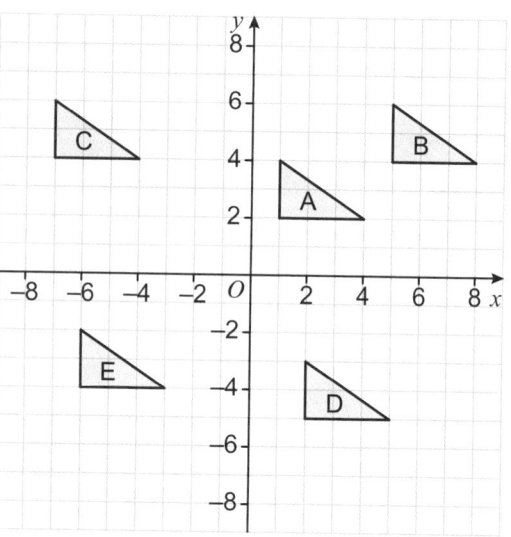

Write down the column vector that maps shape

   a   A onto B
   b   A onto C
   c   A onto D
   d   A onto E
   e   B onto C
   f   B onto D
   g   E onto D.

5   **Exam-style question**

   Describe fully the single transformation that maps parallelogram F onto parallelogram G.

   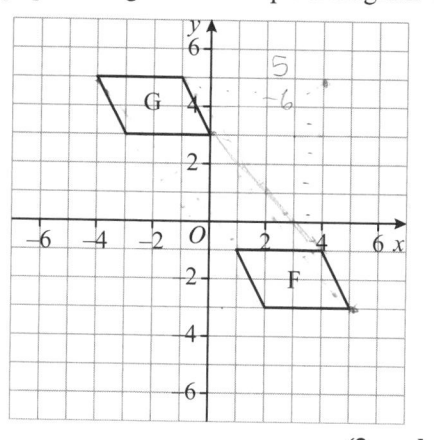

   (2 marks)

Exam hint
The first mark is for the type of transformation; the second mark is for the vector. 'Describe fully' means write the type of transformation (translation) and the vector.

6 Copy the grid and shape A from **Q4**.
  a Translate shape A by the column vector $\begin{pmatrix} 2 \\ -3 \end{pmatrix}$.
    Label the image S.
  b Translate shape S by the column vector $\begin{pmatrix} -7 \\ -2 \end{pmatrix}$.
    Label the image T.
  c Describe fully the transformation that maps
    A onto T.

10.2 Reflection

1 Copy the diagram onto a coordinate grid with
  x- and y-axes from –6 to 6.

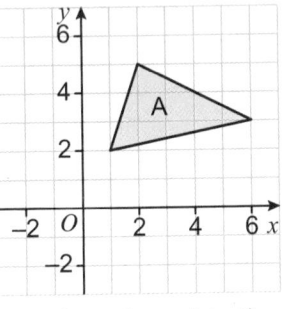

  a Reflect shape A in the x-axis.
    Label the image B.
  b Reflect shape A in the y-axis.
    Label the image C.
  c Reflect shape B in the y-axis.
    Label the image D.

2 Copy the diagram from **Q1** onto a coordinate
  grid with x from –8 to 8 and y from –2 to 6.
  a Draw the line $x = -1$.
  b Reflect shape A in the line $x = -1$.
    Label the image X.
  c Draw the line $y = 2$.
  d Reflect shape A in the line $y = 2$.
    Label the image Y.

3 **R** Copy the diagram onto a coordinate grid
  with x from –6 to 8 and y from –6 to 6.

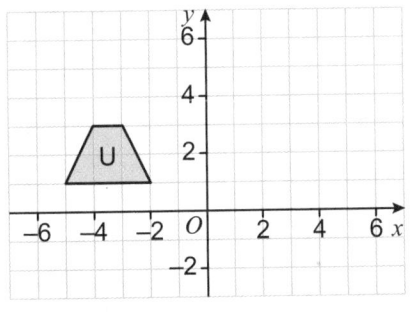

  a Reflect shape U in the line $x = 1$.
    Label the image V.
  b Reflect shape V in the line $y = -1$.
    Label the image W.

c Jack starts with shape U and reflects it in
  $y = -1$ and then in $x = 1$.
  Does he get the same final image?

4 Copy the diagram from **Q3** onto a coordinate
  grid with x from –6 to 8 and y from –6 to 6.
  Reflect shape U in the line $y = 2$.

5 Copy the diagram onto a coordinate grid with
  x and y-axes from –8 to 8.

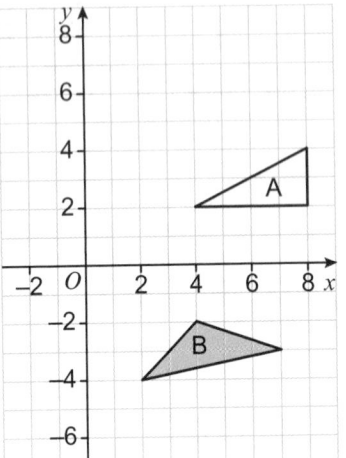

  a Reflect shape A in the line $y = x$.
    Label the image A'.
  b Reflect shape B in the line $y = -x$.
    Label the image B'.

Q5 hint Count the perpendicular distance of
each vertex from the mirror line, then count
the same again the other side of the line.

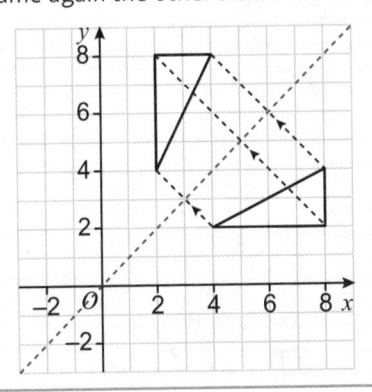

6 **Exam-style question**

Reflect shape A in the line $y = -x$.

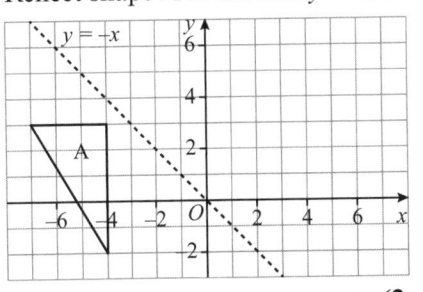

(2 marks)

7 **R** Describe fully the transformation that maps shape
  a A onto B   b A onto C
  c C onto E   d B onto D.

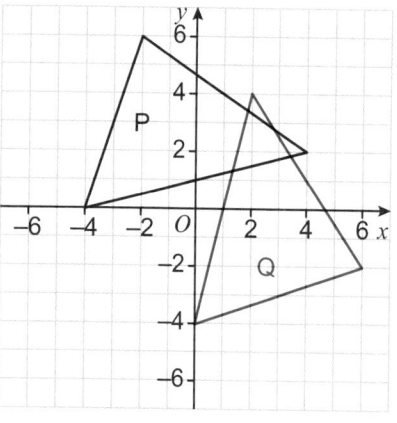

8 **R** Describe fully the transformation that maps shape P onto shape Q.

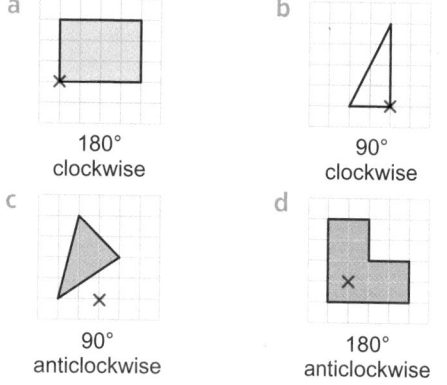

10.3 Rotation

1 Copy each diagram.
  Use the centre of rotation to draw the image of each shape after the rotation given.

  a

  180°
  clockwise

  b

  90°
  clockwise

  c

  90°
  anticlockwise

  d

  180°
  anticlockwise

2 Copy the diagram onto a coordinate grid with x- and y-axes from –8 to 8.

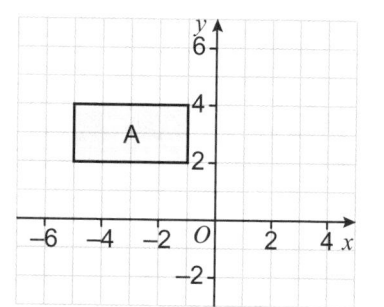

  Rotate shape A 90° clockwise about each centre of rotation.
  a (0, 0)   b (0, –2)
  c (0, 3)   d (–4, 2)

3 Copy the diagram.

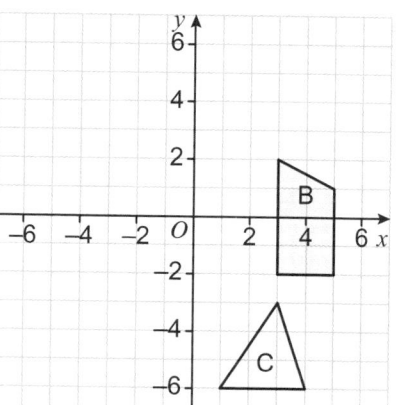

  a Rotate shape B 90° anticlockwise about (0, –4).
     Label the image B'.
  b Rotate shape C 180° clockwise about (1, –1).
     Label the image C'.

4 **R** Shape V can be rotated onto shape W.

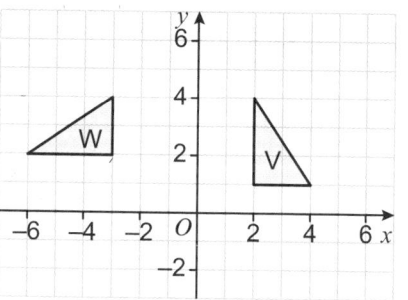

  a What is the direction and what is the angle of rotation?
  b Where is the centre of rotation?
  c Describe fully the transformation that maps shape V onto shape W.

5 **R** Describe fully the transformation that maps shape

a A onto B

b A onto C

c A onto D

d B onto C.

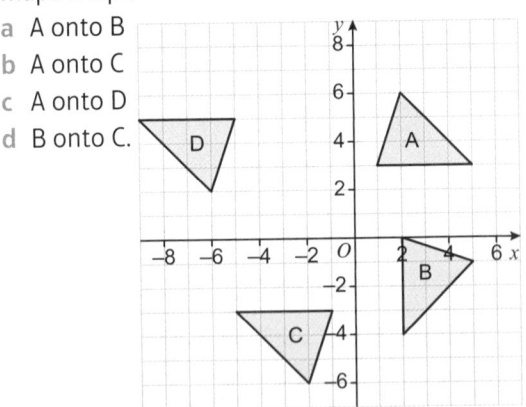

6

Exam-style question

Describe fully the single transformation that maps quadrilateral S onto quadrilateral T.

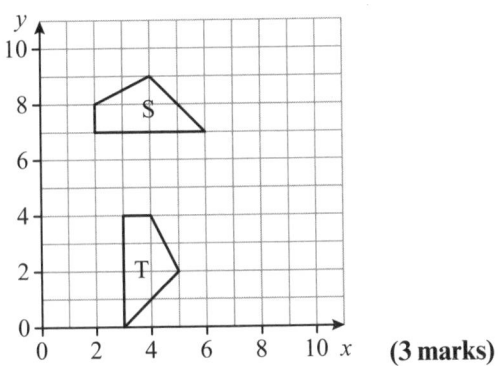

(3 marks)

Exam hint

For 3 marks you must include (1) the one type of transformation, (2) the angle and direction and (3) the coordinates of the centre of rotation.

10.4 Enlargement

1 Copy the diagrams.
Enlarge each shape by the scale factor.

a

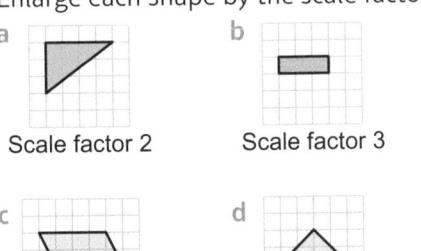

Scale factor 2

b

Scale factor 3

c

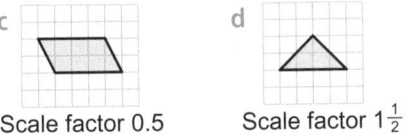

Scale factor 0.5

d

Scale factor $1\frac{1}{2}$

2 Copy each diagram.
Enlarge each shape by the scale factor from the centre of enlargement.

Example

a

b

Scale factor 2   Scale factor 3

c

d

Scale factor 0.5   Scale factor 2

Q2 hint To check your answer, draw lines from the centre of enlargement through the vertices on the original shape and across the grid. These lines should go through the vertices of the image.

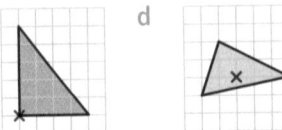

3 Copy the diagram.
Enlarge shape A by scale factor 2 from the given centre of enlargement.

a (0, 4)

b (1, 0)

4 Copy the diagram.
Enlarge each shape by the scale factor from the centre of enlargement.

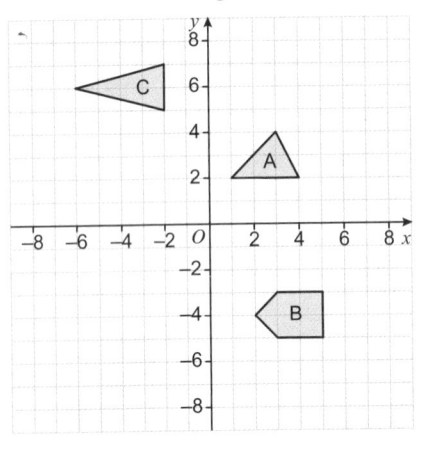

a Shape A   Scale factor 2, centre (0, 1)

b Shape B   Scale factor 3, centre (5, −3)

c Shape C   Scale factor $1\frac{1}{2}$, centre (0, 9)

5 Megan enlarges this shape on her computer.

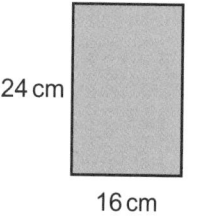

24 cm

16 cm

a She enlarges it by 120%.
Work out the length and width of this enlargement.

b She enlarges the original shape by 80%.
Work out the length and width of this enlargement.

10.5 Describing enlargements

1 Shape B is an enlargement of shape A.

Write down the scale factor of enlargement for each pair of shapes.

Example

a

A

B

b

B

A

2 **R** Triangle PQR is an enlargement of triangle ABC.

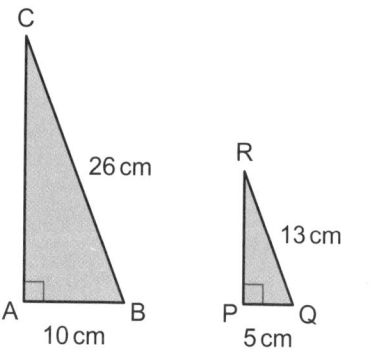

C

26 cm

A   B
10 cm

R

13 cm

P   Q
5 cm

a What is the scale factor of the enlargement?

b AC = 24 cm. Work out the length of PR.

---

3 Shape Q is an enlargement of shape P.

a What is the scale factor of the enlargement?

b What is the centre of enlargement?

c Describe fully the enlargement that maps shape P onto shape Q.

4 **R** Describe fully the transformation that maps shape A onto shape B.

a

b

c

Q4 hint Write: enlargement, scale factor ☐, centre (☐, ☐)

5

Describe fully the single transformation that maps shape J onto shape K.

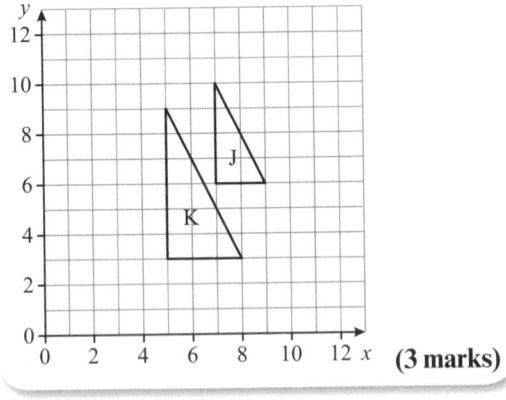

(3 marks)

6 **R** For each diagram

i describe fully the transformation that maps shape A onto shape B

ii describe fully the transformation that maps shape B onto shape A.

a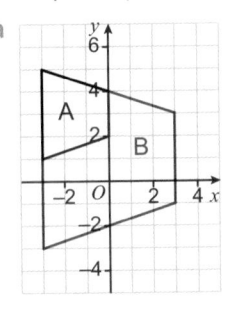

b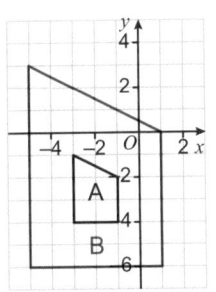

10.6 Combining transformations

1 Copy the grid and shape A.

Example

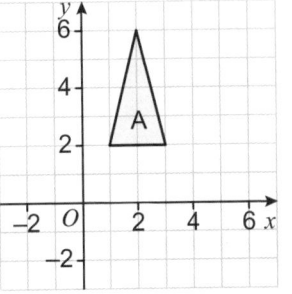

a Reflect shape A in the line $y = x$. Label the image B.

b Reflect shape B in the y-axis. Label the image C.

c Describe fully the single transformation that maps shape A onto shape C.

2 Copy the grid and shape P from **Q1**.

a Reflect shape P in the line $x = 0$. Label the image Q.

b Reflect shape Q in the line $x = 4$. Label the image R.

c Describe fully the single transformation that maps shape P onto shape R.

3 Copy the diagram onto a coordinate grid with x- and y-axes from −6 to 6.

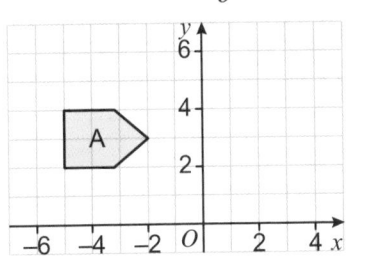

a Reflect shape A in the y-axis. Label the image B.

b Rotate shape B 180° about the origin (0, 0). Label the image C.

c Describe fully the single transformation that maps shape A onto shape C.

4 Copy the diagram onto a coordinate grid with x from −8 to 8 and y from 0 to 8.

a Enlarge shape A by scale factor 2, centre (−6, 0). Label the image B.

b Enlarge shape B by scale factor 0.5, centre (8, 4). Label the image C.

c Describe fully the single transformation that maps shape A onto shape C.

5

┌─────────────────────────────┐
│ **Exam-style question**   │
└─────────────────────────────┘

Shape J is rotated 180° about the origin to give shape K.

Shape K is translated by the vector $\begin{pmatrix} -2 \\ 4 \end{pmatrix}$ to give shape L.

Describe fully the single transformation that maps shape J to shape L.

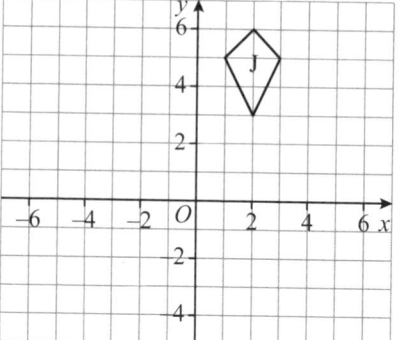

(3 marks)

┌──────────────────────────────────────┐
│ **Q5 hint** Draw the transformations on the │
│ grid. The origin is the point (0, 0). Remember │
│ to give only one transformation. │
└──────────────────────────────────────┘

6 **P** Which single transformation is the same as

a two translations

b a rotation and a translation?

10 Problem-solving

Solve problems using these strategies where appropriate:

- Use pictures
- Use bar models
- Use flow diagrams.
- Use smaller numbers
- Use x for the unknown

1

┌─────────────────────────────┐
│ **Exam-style question**   │
└─────────────────────────────┘

A rectangle has a length of 10 cm and a width of 2 cm.

The rectangle is enlarged.

The enlarged rectangle has a width of 12 cm.

10 cm
2 cm [   ] 12 cm
Enlarged rectangle

Diagram NOT
accurately drawn

Work out the length of the enlarged rectangle. **(3 marks)**

2 The histogram shows the amounts people spend on rent each week.

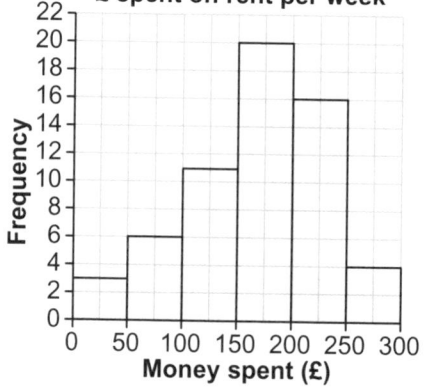

What fraction of these people spend more than £150 per week on rent?

3 **R** This cube has surface area 541.5 cm².

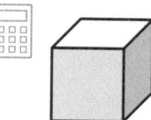

What is the length of one side of the cube?

┌──────────────────────────────────────┐
│ **Q3 hint** Draw a flow diagram. │
└──────────────────────────────────────┘

4 Kevin is creating a stencil to decorate his room. He uses a grid with x from −10 to 5 and y from −5 to 5 and draws a design in one of the quadrants.

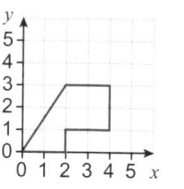

He reflects the design in the line $y = 0$. Kevin then reflects the whole design in the line $x = -1$.

Use a grid to draw Kevin's finished design.

5 Liz asked 300 people at a conference whether they came from the local area, from the UK or from overseas.

Here is some information about her results.

70 of the 140 males were from the UK.

25% of the females were from the local area.

Half of all the females were from overseas.

$\frac{1}{3}$ of the total asked were from the local area.

Work out the total number of people at the conference who were from the UK.

6 Ellen is creating a cartoon. She is making the buildings appear as if they are closer.
For her drawings, Ellen uses a scale factor of 2.

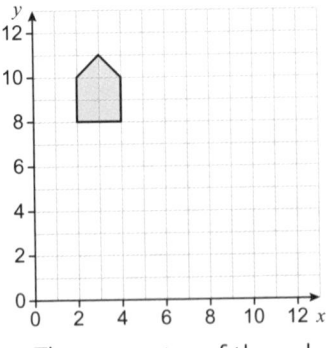

a The top vertex of the enlarged shape is at point (6, 10).
Draw the rest of the enlarged shape.
b What is the centre of enlargement?
c Using the same centre of enlargement, draw the next shape at a scale factor of 3 of the original drawing.

7 Chris has moved the desk in his office. The diagram shows the desk's old position.
Chris rotated the desk 90° clockwise around the origin and translated it $\begin{pmatrix} -4 \\ -1 \end{pmatrix}$.
Copy the diagram and show the new position of the desk.

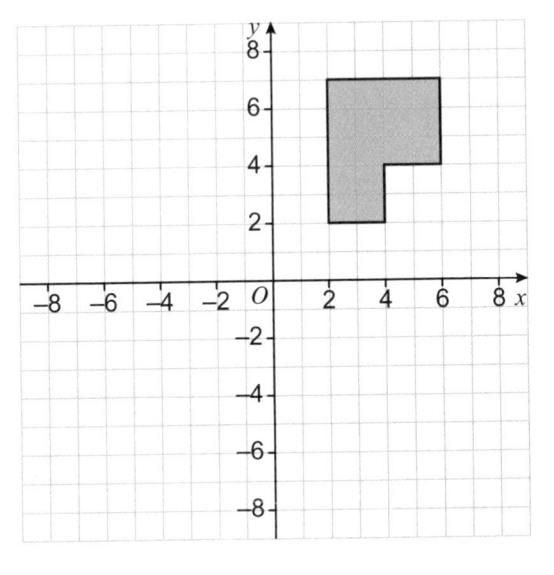

8 **R** Fabien says that two reflections can put a shape in the same place as a single translation.
He uses a square to demonstrate.

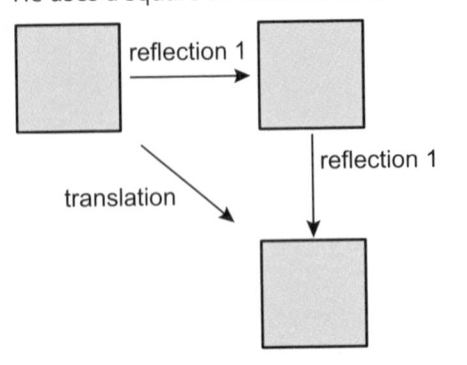

a Crystal is not sure about this and uses an arrow shape to test Fabien's theory.
What did Crystal find?
b Suggest a single transformation that will put Crystal's shape in the same place as two reflections.

9 Chie is creating a map for her robot. The robot starts at position A but must stop at two places (B and C) on the map. Chie writes down the column vectors for the two translations:
$\begin{pmatrix} -4 \\ 2 \end{pmatrix}$ for A to B and $\begin{pmatrix} 0 \\ -7 \end{pmatrix}$ for B to C.
The diagram shows the robot at position C. The top left vertex of the robot is at (−2, −3).

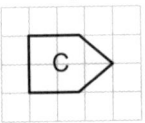

Draw Chie's map on a coordinate grid to show the robot in positions A, B and C.

10 Sarah is creating a design for a sequin motif on a scarf.
She has drawn a grid for her pattern and marked where to put the sequins. She has sequins at coordinates (6, 6), (5, 5), (4, 4), (4, 5), (6, 5), (6, 3), (3, 2), (3, 6) and (2, 4).
She wants to reflect the pattern in the line $y = -x$.
Draw the completed pattern.

11 RATIO AND PROPORTION

11.1 Writing ratios

1 Write down each ratio of blue cubes to yellow cubes.

a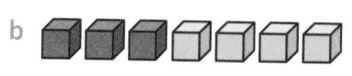

b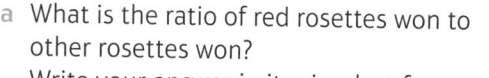

2 Draw cubes to show these ratios of blue to yellow.
   a 5 : 2   b 2 : 3

3 A box of chocolates has 20 chocolates. There is 1 nut chocolate for every 4 caramel chocolates.
   How many chocolates are
   a nut
   b caramel?

4 Write each ratio in its simplest form.
   a 3 : 15   b 12 : 6
   c 32 : 8   d 6 : 48
   e 15 : 25   f 72 : 48
   g 28 : 49   h 36 : 42

   Q4a hint

   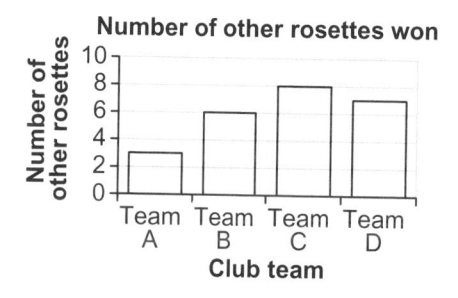

   $$3 : 15$$
   $$÷3 \quad\quad ÷3$$
   $$1 : \square$$

5 **P / R** The bar charts show the numbers of rosettes won in a horse show by teams from a club.

   Number of red rosettes won

   [Bar chart: Number of red rosettes (y-axis 0–5) vs Club team. Team A = 3, Team B = 0, Team C = 4, Team D = 2]

   Number of other rosettes won

   [Bar chart: Number of other rosettes (y-axis 0–10) vs Club team. Team A = 3, Team B = 6, Team C = 8, Team D = 7]

a What is the ratio of red rosettes won to other rosettes won?
   Write your answer in its simplest form.

b The club's trainer says that they won three times as many other rosettes as red rosettes.
   Is the trainer correct?
   Explain your answer.

6 A primary school class teacher is planning a class trip. There are 30 children in the class and 5 adults are available to go on the trip. The guidelines say that the adult-to-child ratio should be 1 : 6.
   Are there enough adults for the trip?

7 Which of these ratios are equivalent?
   A 56 : 49   **B** 120 : 105
   C 108 : 84   **D** 128 : 112
   E 63 : 49

8 Write each ratio in its simplest form.
   a 6 : 21 : 15
   b 15 : 25 : 30
   c 32 : 40 : 48
   d 45 : 72 : 54

9 Show that these ratios are equivalent.
   15 : 10 : 35   18 : 12 : 42   36 : 24 : 84

10 A recipe for sweet pastry uses 200 g of flour, 100 g of butter and 75 g of sugar.
   Write the ratio of flour : butter : sugar in its simplest form.

11 **Exam-style question**

   There are 60 sweets in a bag. Of these, 18 are toffee and the rest are fruit flavoured. Write the ratio of toffee to fruit-flavoured sweets in its simplest form. **(2 marks)**

11.2 Using ratios 1

1 The ratio of sand : cement needed to make mortar is 5 : 1.
   Sarah uses 6 kg of cement.
   How many kilograms of sand does she use?

   Example

2 Mark makes a model of the Burj Khalifa tower using a ratio of 1 : 2400.
   The height of his model is 345 mm.
   What is the height of the Burj Khalifa tower in metres?

   Q2 hint 1 : 2400 means that 1 mm in the model represents 2400 mm in real life.

3 Write each ratio as a whole number ratio in its simplest form.

   Example

   a 0.3 : 6
   b 3.2 : 4.8
   c 36 : 1.5
   d 14.4 : 25.6

4 Write each ratio as a whole number ratio in its simplest form.
   a 0.3 : 1.25
   b 37.8 : 6.3
   c 12.5 : 13.75
   d 28.8 : 61.2

5 **P** Windows are made in standard ratios.
   Type A are made in the width : height ratio of 5 : 6, and Type B are made in the ratio 5 : 8.
   Is each of these windows Type A or Type B?

   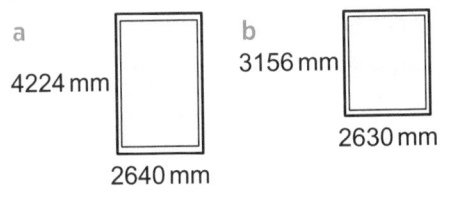

   a
   4224 mm
   2640 mm

   b
   3156 mm
   2630 mm

6 Brass screws are made from copper and zinc in the ratio 84.5 : 45.5.
   Write this ratio in its simplest form.

7 The ratio of fat : flour in a recipe is 3 : 8.
   Aziz uses 75 g of fat.
   How many grams of flour should he use?

8 **R** The ratio of potato : cauliflower in a potato and cauliflower curry recipe is 9 : 15.
   Sanjay uses 420 g of cauliflower.
   He has 300 g of potato.
   Will he use all of the potato?

9 Huw splits his weekly pocket money into spending money and money saved to buy a laptop in the ratio 17 : 13.
   He gets £15 each week.
   How much money does he save each week?

10 **Exam-style question**

   A school organises a baking competition. Four classes take part. Each student can only enter one item in the competition.

   One class enters cakes and loaves of bread in the ratio 3 : 1.

   There are 6 loaves of bread entered in the competition.

   All the entries from the other classes are cakes.

   The same number of students take part from each class.

   How many students take part in the competition altogether?   **(4 marks)**

11.3 **Ratios and measures**

1 Write these ratios in their simplest form.
   a 1 hour : 20 minutes
   b 750 ml : 1 litre
   c 8 mm : 2 m

   Q1 hint Both parts of the ratio need to be in the same units before you simplify.

2 A can contains 250 ml of cola and a bottle contains 2 litres of cola.
   Write down the ratio of the amount of cola in the can to the amount of cola in the bottle. Give your answer in its simplest form.

3 Complete these conversions.

   Example

   a 64 000 g = ☐ kg
   b 4.7 litres = ☐ ml
   c 135 mm = ☐ cm
   d 3.6 km = ☐ m

4 1 mile ≈ 1.6 km. Convert
   a 8 km to miles
   b 62 miles to km.
   c The distance from Paris to Brussels by car is 305 km. How far is this in miles?

5 A recipe needs $1\frac{1}{2}$ pints of milk.
   a 1 pint = 20 fl. oz (fluid ounces).
      Convert $1\frac{1}{2}$ pints to fluid ounces.
   b 1 fl. oz ≈ 28 ml. Convert your answer to part **a** into millilitres.

6

Exam-style question

A farm shop has 30 hundredweight of potatoes.

They need to be put into 5 kg bags to be sold.

The farm shop manager knows that
- 1 hundredweight = 112 pounds
- 1 pound ≈ 0.454 kilogram

How many bags of potatoes will the farm shop have for sale? **(4 marks)**

7 On the day that David buys some euros for his holiday, £1 buys €1.27.
How many euros does he get for £400?

8 Choose an amount in £ from the cloud.
Choose an exchange rate from the table.

£300 £425
£280 £550

Currency	£1
Euros	1.27
Turkish lira	3.56
Croatian kuna	9.71
Albanian lek	176.22
Tunisian dinar	2.90
Moroccan dirham	14.06

Work out how much money you will get.
Repeat for two more amounts and currencies.

9 **P** Use the table in **Q8**.
a How many £ does Vicky get for 500 Turkish lira?
b Which is worth more, 500 Moroccan dirham or 3800 Croatian kuna?

10 **P** The diagram shows two right-angled triangles.

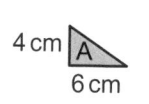
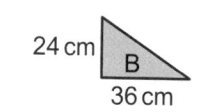

4 cm A
6 cm

24 cm B
36 cm

Write these ratios in their simplest form.
a The base of triangle A to triangle B.
b The height of triangle A to triangle B.
c The area of triangle A to triangle B.

11 The diagram shows two cuboids.

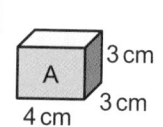
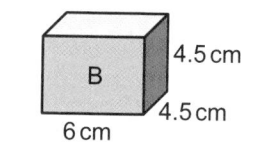

A 3 cm
3 cm
4 cm

B 4.5 cm
4.5 cm
6 cm

Write these ratios in their simplest form.
a Height of A to height of B.
b Area of the darker shaded face of A to area of the darker shaded face of B.
c Volume of A to volume of B.

12 A pencil case contains just 4 pens and 9 pencils.
a What is the ratio of pens : pencils?
b What fraction of the contents of the pencil case are pens?
c What fraction of the contents of the pencil case are pencils?

13 In a multipack of crisps, $\frac{3}{5}$ of the packets of crisps are salt and vinegar and $\frac{2}{5}$ of the packets are cheese and onion.
a What is the ratio of salt and vinegar : cheese and onion?
b There are 6 packets of cheese and onion in the multipack. How many packets of salt and vinegar are there?
c How many packets of crisps are in the multipack altogether?

14 **R** In a school drama group, $\frac{3}{4}$ of the group are girls and the rest are boys.
a What is the ratio of boys : girls in the group?
b Rashid thinks there are 27 students in the group. Explain why he must be wrong.

11.4 Using ratios 2

1 Share these amounts in the ratios given.
a 24 kg in the ratio 3 : 1
b 72 m in the ratio 1 : 5
c £32 in the ratio 5 : 3
d £250 in the ratio 3 : 7
e 42 litres in the ratio 5 : 2
f 12.5 km in the ratio 3 : 2

Example

2 Solder used for joining metal pipes is made from a mixture of tin and lead in the ratio 3 : 2. Paul has 425 g of solder.
What is the mass of
a tin   b lead?

3 Green dye is mixed from yellow dye and blue dye in the ratio 3 : 4.
Amir needs 35 litres of green dye.
a How many litres of yellow dye should he use?
b How many litres of blue dye should he use?

4 A cereal mix is made from oats, wheat and barley in the ratio 3 : 2 : 1.
Chantal wants to make 600 g of cereal mix. How much of each type of ingredient does she need?

5 **Exam-style question**

Jamila is going to make some shortbread.

She needs to mix flour, butter and sugar in the ratio 3 : 2 : 1 by weight.

Jamila wants to make 1.5 kg of shortbread. She has

- 450 g of butter
- 1500 g of flour
- 500 g of sugar

Does Jamila have enough flour, butter and sugar to make the shortbread? **(4 marks)**

6 Share these amounts in the ratios given.
a 750 g in the ratio 2 : 3 : 5
b 600 ml in the ratio 3 : 2 : 7
c £84 in the ratio 3 : 5 : 6
d £30 in the ratio 4 : 5 : 3

7 Share these amounts in the ratios given. Round your answers sensibly.
a £60 in the ratio 3 : 4
b 125 g in the ratio 5 : 4

8 **R** Jennifer and Tom buy a flat to rent out for £125 000.
Jennifer pays £75 000 and Tom pays £50 000. They rent out the flat for £650 per month.
a Write the amounts they each pay as a ratio.
b Write the ratio in its simplest form.
c Divide the rent in this ratio. How much of the rent does each of them get?

Q8b hint

Jennifer : Tom

÷ □ (□ : □ / □ : □) ÷ □

9 **P** Kamal and Aisha buy a car together for £6300.
Aisha pays £2700 and Kamal pays £3600. They sell the car 4 years later for £4200.
How should they share the money fairly?

10 **P / R** Two whole numbers are in the ratio 2 : 5 and their product is a multiple of 3.
What are the smallest numbers possible?

11 Sharon and Jo share £90 in the ratio 7 : 5.
a How much do they each receive?
b How much more does Sharon get than Jo?

12 **Exam-style question**

Kevin and Lesley share a bag of sweets in the ratio 3 : 4

Kevin gets 15 fewer sweets than Lesley.

How many sweets did Lesley get? **(3 marks)**

11.5 **Comparing using ratios**

1 In a karate class $\frac{4}{9}$ of the group are men and the rest are women.
What is the ratio of men to women in the class?

2 Copy and complete the table for different groups of people having karate lessons.

Fraction of group that are men	Ratio of men : women
$\frac{3}{7}$	
	5 : 3
$\frac{11}{12}$	
	5 : 6

3 In a particular fruit juice mix, $\frac{5}{12}$ is orange juice, $\frac{4}{12}$ is mango juice and the rest is apple juice.
Work out the ratio of
orange juice : mango juice : apple juice.

4 A car windscreen washer mix can be made from water and concentrate in the ratio 5 : 1.
What fraction of the washer mix is
a water
b concentrate?

5 Mark and Jon share a pizza in the ratio 5 : 4.
What fraction of the pizza should
  a Jon get
  b Mark get?

6 Gilding brass is made of copper and zinc in the ratio 19 : 1.
What fraction of the brass is
  a copper
  b zinc?

7 **R** The ratio of boys to girls in a music club is 3 : 7.
Curtis says that $\frac{3}{7}$ of the club members are boys.
Is he correct? Explain your answer.

8 **R** Fruit cookies are made by mixing flour, butter, sugar and dried fruit in the ratio 3 : 2 : 1 : 1.
  a What fraction of the cookie mix is
    i flour
    ii butter?
  b In 700 g of cookie mix, what are the masses of flour, butter, sugar and dried fruit?
  c Joyce has plenty of flour and dried fruit, but only 150 g of butter and 200 g of sugar.
  What is the maximum amount of cookie mix she can make?
  Show your working clearly.

9 **R** A building mix uses cement, lime, building sand and gravel in the ratio 1.5 : 1.5 : 5 : 2
  a What fraction of the mix is lime?
  b In 50 kg of the mix, what are the masses of cement, lime, building sand and gravel?
  c You have plenty of cement, gravel and lime, but only 22.5 kg of building sand.
  What is the maximum amount of the building mix you can make?

> **Q9 hint** Write the ratio with whole numbers first.

10 Copy and complete to write these as unit ratios.
  a
  $$\div 5 \overset{7 : 5}{\underset{\square : 1}{\frown}} \div 5$$
  b
  $$\div 4 \overset{4 : 9}{\underset{\square : \square}{\frown}} \div 4$$

11 Write each of these in the form $m : 1$.
Give each answer to a maximum of 2 decimal places.
  a 11 : 4
  b 17 : 3
  c 56 : 15
  d 3 : 16
  e 7 : 32
  f 23 : 60

12 **R** Millie and Hannah are dyeing T-shirts.
Millie makes an orange dye by mixing 75 g of yellow dye with 25 g of red dye.
Hannah mixes 60 g of yellow dye with 24 g of red dye for her orange dye.
Whose dye is darker orange?
Explain your answer.

Example

13 **R** Liam makes wallpaper paste by mixing 230 g of glue with 8 litres of water.
Ryan makes some more wallpaper paste by mixing 95 g of glue with 3500 ml of water.
Who makes the stronger paste?
Explain your answer.

14 **R** Nadya makes a recipe that uses flour to butter in the ratio 500 : 190.
Karl makes a different recipe using flour to butter in the ratio 425 : 150.
Whose recipe has the higher proportion of butter?

15 **R** Tina and Gareth are running a marathon.
Tina runs for 4 hours 45 minutes and walks for 45 minutes.
Gareth runs for 5 hours 15 minutes and walks for 75 minutes.
Who walks for a higher proportion of their race time?
Explain your answer.

16
> **Exam-style question**
>
> A paint mixture is $\frac{4}{7}$ blue and $\frac{3}{7}$ red.
> Jack says, 'The amount of blue is one and a third times the amount of red.'
> Is he correct? Explain your answer.
>
> **(3 marks)**

11.6 Using proportion

1 6 apples cost £1.50.
   What is the cost of
   a 12 apples   b 2 apples
   c 4 apples   d 10 apples?

Example

2 4 tickets to a show cost £80.
   How much will 17 tickets cost?

3 60 litres of compost cost £4.50.
   What do 84 litres cost?

4 Edward gets £43.05 for 7 hours of work.
   How much will he get for working for
   a 12 hours   b 25 hours
   c 35 hours?

5 **R** A group of 14 friends want to go on holiday
   together. The total cost of the holiday is
   £2996 and they all pay the same amount.
   3 more friends agree to join them.
   What is the total cost of the holiday now?

6 **R** Ketchup bottles come in two sizes.

   342 g for £1.25   460 g for £1.75

   a Copy and complete to find the unit ratios.

   i

   342 g : £1.25
   ÷ □   ÷ □
   □ g : £1

   ii
   460 g : £1.75
   ÷ □   ÷ □
   □ g : £1

   b Which size is the better buy?

   Q6b hint Which gives more grams for £1?

7 **R** Kasia can buy bottles of cola in two sizes:
   a 1.25-litre bottle for £1.55 or a 1.75-litre
   bottle for £1.85.
   Which is better value for money?

8 **R** Ajay can buy bags of rice in two sizes:
   1.5 kg for £3.27 or 600 g for £1.32.
   Which is better value for money?

9 **R** Tom and Fred are both house painters.
   Tom charges £146.10 labour for a job that
   takes 6 hours. Fred charges £262.90 labour
   for a job that takes 11 hours.
   Who will earn more for a job that takes
   17 hours?

10 **Exam-style question**

   Jill is making cupcakes. She finds out
   the prices of cupcake cases from two
   companies, Bake Rite and Kitchen Plus.

Kitchen Plus	Bake Rite
Pack of 60 cupcake cases £2.25	Pack of 80 cupcake cases £2.89

   She needs 240 cupcake cases and wants to
   buy them as cheaply as possible.

   Which company should Jill buy the
   cupcake cases from?

   You must explain your answer. **(4 marks)**

11.7 Proportion and graphs

1 Look at the graph.

Conversion graph

(y-axis: Pounds, x-axis: Kilograms)

   Are kilograms and pounds in direct proportion?
   Explain your answer.

2 The table shows some amounts of liquid in
   litres and in pints.

Litres	2	4	6	10
Pints	3.5	7	10.5	17.5

   a Plot a line graph for these values.
   b Are litres and pints in direct proportion?
      Explain your answer.
   c Car oil is sold in a 5-litre can.
      How many pints is this?

3 The table shows the price of nails at a DIY store.

Mass of nails (kg)	0.5	1	2	5
Price of nails (£)	1.20	2.40	4.80	12

a Plot a line graph for these values.

b Is the price of the nails in direct proportion to the mass? Explain your answer.

4 This line graph shows the price of petrol by volume.

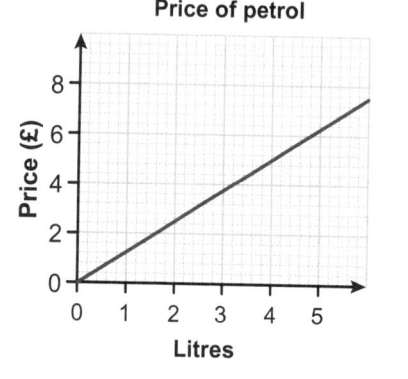

Price of petrol

a Are price and volume in direct proportion?

b Work out the gradient of the line.

c How much does 1 litre of petrol cost?

5 **R** Which of these are in direct proportion? Give reasons for your answers.

a Miles and kilometres

b Age and number of pets

c Litres and millilitres

d Number of people at a bus stop and time of day

6 **P** An electrician charges a callout fee of £35 plus £42 per hour she works.

Is her total charge (C) in direct proportion to the number of hours (h) she works?

7 Which of these sketch graphs show one variable in direct proportion to the other?

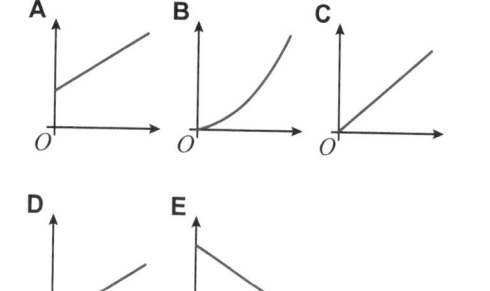

8 The table shows the amount paid for different size containers of milk.

Number of litres, n	0.6	1	1.2	2	2.4
Cost, C (p)	48	80	96	160	192

a Draw the graph.

b Work out the equation of the line.

c Are C and n in direct proportion?

d Write a formula linking the number of litres (n) and the cost (C).

e Use the formula to work out the cost of 3.6 litres of milk.

9 Look back at the graph in **Q1**.

a Write the ratio kilograms : pounds in its simplest form.

b Work out the gradient of the graph.

c Write a formula that links kilograms (K) and pounds (P).

10 On a particular day, the ratio of GB pounds to US dollars is 5 : 9.

a Copy and complete this table of values for pounds and dollars.

GB pounds (£)	0	5	10
US dollars ($)			

b Draw a conversion graph for pounds to dollars.

c Write 5 : 9 as a unit ratio.

d Write a formula linking dollars (y) and pounds (x).

11
Exam-style question

A plumber uses a formula to work out the cost (C) of a job according to the number of hours (H) it takes. His formula is

$C = 35H + 45$

Is the plumber's charge for a job in direct proportion to the amount of time it takes? Explain your answer. **(2 marks)**

Exam hint

You won't get any marks for just 'Yes' or 'No'. You must write 'Yes' or 'No' because …………..

11.8 Proportion problems

1 3 children build a giant sandcastle in 4 hours. How long will it take
   a 1 child
   b 5 children?

Example

> **Q1a hint** Will it take 1 child more or less time than 3 children?

2 2 decorators take $5\frac{1}{2}$ hours to paint 3 rooms. How long will it take
   a 3 decorators
   b 5 decorators?
   Give your answers in hours and minutes.

3 3 school uniform shirts cost £13.50. How much will 5 shirts cost?

4 **R** A farmer needs to pick his 750 kg crop of strawberries.
   He wants all the work done in 8 hours or less.
   He knows that 1 person can pick 8 kg of strawberries in an hour.
   How many people does he need to do the job?

5 A group of 5 cyclists can complete a circular 20-mile cycle in 2 hours.
   How long will it take a group of 8 cyclists to complete the same ride?

6 It takes 25 minutes for 24 cupcakes to cool. How long will it take for 32 cupcakes?

7 **R** A farmer has enough bales of silage to feed 120 cattle for 30 days.
   He sells 30 cattle.
   How many days will the silage last for?

8 A print run of 40 exam papers uses 560 sheets of paper.
   How many sheets of paper would be needed to print 215 copies of the same exam paper?

9 **R** A caretaker needs to clean 35 classrooms.
   He knows that 2 people can clean a room in 20 minutes.
   He needs all the work to be done in 3 hours after school.
   How many cleaners does he need to employ?

10 **R** It costs £85.82 to hire a car for 7 days.
   There is no standing charge.
   How many days does Amit hire a car for if he pays £208.42?

11 **R** It takes 6 people 5 hours to pick 600 kg of apples.
   The apple grower employs 2 more people.
   How long will it take to pick 600 kg of apples now?

12 (**Exam-style question**

   It takes 7 men 4 hours to dig a ditch.
   How long would it take 5 men? **(2 marks)**

11 Problem-solving

Solve problems using these strategies where appropriate:
- **Use pictures**
- **Use smaller numbers**
- **Use bar models**
- **Use x for the unknown**
- **Use a flow diagram**
- **Use more bar models.**

Example

1 Jay, Alix and Cam have dinner.
   The total bill is £40.
   Jay has only a starter. He pays one tenth of the bill.
   Cam has extra drinks. She pays half of the bill.
   a How much does Jay pay?
   b How much does Alix pay?
   c What is the ratio of Alix's bill to Cam's bill?
     Write your answer in its simplest form.

2 Ajesh has 8 box sets of his favourite drama.
   He stacks them in a pile next to 20 DVDs.
   Both piles are the same height.
   a What is the ratio of the number of box sets to DVDs?
     Write your answer in its simplest form.
   b One DVD is about 1.4 cm thick.
     How thick is each box set?

3 A group of 6 students takes 3 days to paint a classroom.
   A group of 4 students takes 4 days to paint an identical classroom.
   Do the students in the first group work faster than the students in the second group?

4 **R** Ms Chapman wants to hire minibuses for a Year 8 trip to the museum.
The teacher : student ratio must be at least 1 : 12 on each minibus.
60 students are going on the trip.
Each minibus seats 26 people.
  a What is the lowest number of teachers needed?
  b How many minibuses will be needed?
  c How many spare seats will there be?

5 **R** 3 bags, A, B and C, contain different amounts of money.
The amount in bag B is $\frac{3}{5}$ of the amount in bag A. The amount in bag C is $\frac{1}{4}$ of the amount in bag B.
  a There is £60 in bag A.
    How much money is in bags B and C?
  b What is the ratio A : C in its simplest form?
  c Would the ratio be different if bag A contained £80? Explain your answer.

6 **R** All the sweets in a packet are either cubes or spheres.
The ratio of cubes to spheres is 3 : 5.
  a How many sweets could be in the bag? What is the smallest number possible?
The sweets have either blue wrappers or yellow wrappers.
The ratio of blue to yellow wrappers is 5 : 7.
  b What is the smallest number of sweets possible now?
  c With the number of sweets found in part **b**, how many of each type are there?

> **Q6b hint** You need a number that works for both ratios. Draw bar models to help you.

7 **R** Two of the angles in a quadrilateral are 60° and 90°.
The other two angles are in the ratio 3 : 4.
  a What are the missing angles?
  b There are two different types of quadrilateral with these angles.
    What are they?

8 The first term in a sequence is n.
The term-to-term rule is ×4 − 2
The fourth term is 22.
What is n?

9 **R** Kirti places a 1 kg bag of sugar on her faulty scales. It reads 1.2 kg. She then places a 5 kg bag of rice on it. It reads 6 kg.
  a What is the ratio of real weight to faulty reading of the scales?
  b A recipe requires 800 g of flour.
    What do the scales need to show for Kirti to weigh out the correct amount?

10 Last year, Devan, Nikhil and Jasdeep shared their company profits in the following way.
Devan got $\frac{9}{20}$, Nikhil got $\frac{3}{10}$ and Jasdeep got the rest.
This year they are splitting their profits in the ratio 5 : 4 : 3.
In both years the company's profit was £180 840.
  a Nikhil is happy with the change. Explain why.
  b Using fractions, show that the change did not make a difference to Jasdeep.

11
> **Exam-style question**
>
> At St Andrew's High there are 160 Year 11 students.
>
> 65% of them will be staying on to attend the sixth form college.
>
> Of those staying on, the ratio of boys to girls is 6 : 7
>
> Of those leaving at the end of the year, the ratio of boys to girls is 5 : 2
>
> How many more boys are staying on than leaving? **(5 marks)**

12 RIGHT-ANGLED TRIANGLES

12.1 Pythagoras' theorem 1

1 a Draw each triangle on centimetre squared paper.

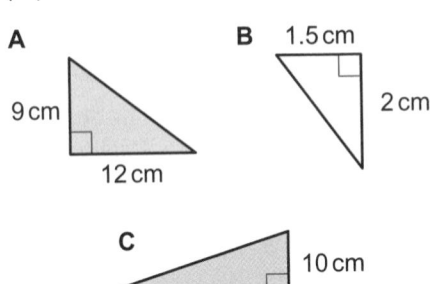

A
9 cm, 12 cm

B
1.5 cm, 2 cm

C
10 cm, 24 cm

b i Measure the length of the unknown side.

ii What is the length of the hypotenuse?

2 Write the length of the hypotenuse for each of these triangles.

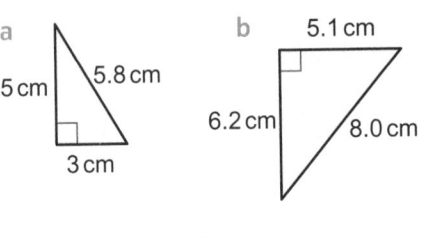

a 5 cm, 5.8 cm, 3 cm

b 5.1 cm, 6.2 cm, 8.0 cm

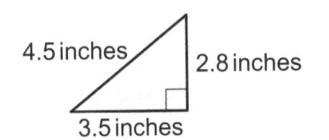

c 4.5 inches, 2.8 inches, 3.5 inches

3 A square is drawn on each side of a right-angled triangle.

a Check that the measurements match your answer to **Q1a**.

b Find the area of each square.

c Copy and complete:
$$\Box^2 = 9^2 + 12^2$$

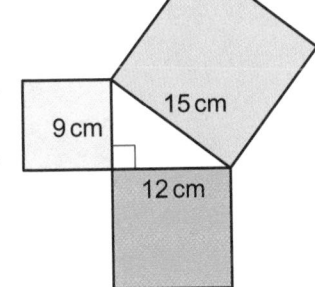

15 cm, 9 cm, 12 cm

4 R Copy and complete this statement.
$$c^2 = \Box^2 + b^2$$

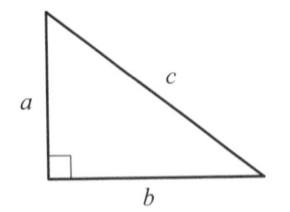

5 Calculate the length of the hypotenuse, x, in each right-angled triangle. Give your answers correct to 3 s.f.

Example

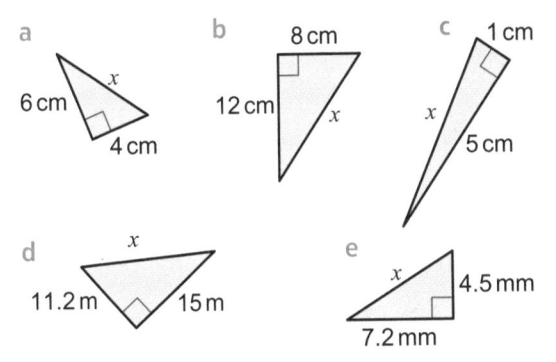

a 6 cm, x, 4 cm

b 8 cm, 12 cm, x

c 1 cm, x, 5 cm

d x, 11.2 m, 15 m

e x, 4.5 mm, 7.2 mm

6 Calculate the length of BC in each right-angled triangle. Give your answers correct to 2 d.p.

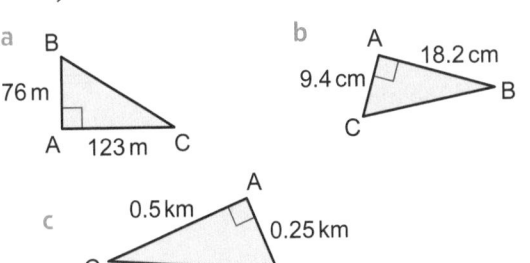

a B, 76 m, A, 123 m, C

b A, 18.2 cm, 9.4 cm, B, C

c A, 0.5 km, 0.25 km, C, B

7 P A ramp is to be installed in a skate park. The ramp will rise 0.5 m over a distance of 4 m and forms a right-angled triangle. How long must the ramp be? Give your answer to the nearest cm.

8

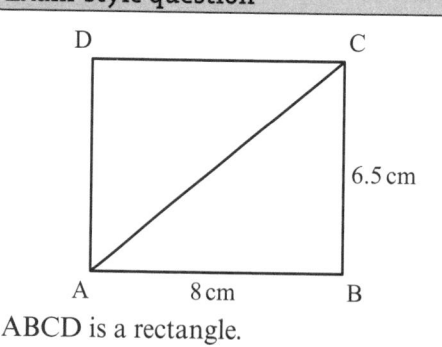

ABCD is a rectangle.

Work out the length of AC.

Give your answer to 3 significant figures. **(4 marks)**

9

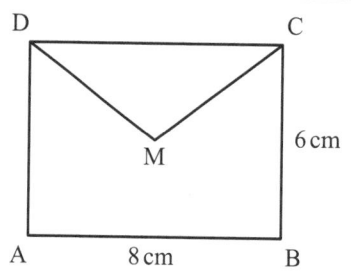

An envelope measures 6 cm by 8 cm

The flap meets the envelope at the point M.

The point M is in the exact centre of the envelope.

A glue strip runs from D to M and M to C.

What is the length of the glue strip? **(4 marks)**

12.2 Pythagoras' theorem 2

1 Are these numbers surds?

a $\sqrt{100}$ b $\sqrt{5}$

c $\sqrt{1}$ d $\sqrt{7}$

e $\sqrt{144}$ f $\sqrt{10}$

2 Simplify these expressions. Give each answer as a surd.

a $\sqrt{2^2 + 5^2}$

b $\sqrt{1^2 + 3^2}$

c $\sqrt{4^2 - 2^2}$

3 **P** The points A(1, 7) and B(4, 2) are shown on a centimetre square grid.

Work out the length in centimetres of AB. Give your answer correct to 3 s.f.

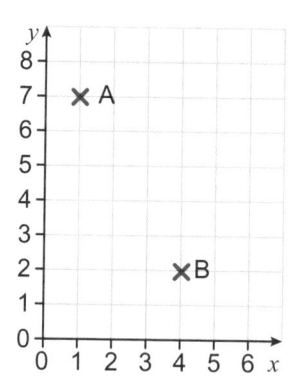

4 **P** A helicopter flies from a lighthouse to a ship. Calculate the distance it flies.

The points are marked on a mile square grid.

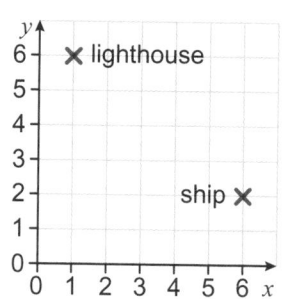

5 The points A and B are plotted on a centimetre square grid.

For each set of points, calculate the length of AB.

a A(1, 1) and B(9, 2)

b A(3, 8) and B(−2, 5)

c A(10, −4) and B(3, 0)

6

ABC is a right-angled triangle.

Example

AB = 12.5 cm and BC = 15.7 cm.

Calculate the length of AC, giving your answer to 3 significant figures. **(3 marks)**

7 Work out the length of the unknown side in each right-angled triangle.
Give your answers to an appropriate degree of accuracy.

a
3.1 m, 5.4 m, x

b
10.2 km, 6.76 km, x

8 **P** PQRS is a square. The diagonal PR = 12 cm.
Work out the length of each side of the square.

P   Q
12 cm
S   R

9 Find the length of the unknown side in each right-angled triangle.
Give your answers in surd form.

a
B, 12 cm, C, x, A, 10 cm

b
A, 5 cm, B, x, 9 cm, C

c
2 cm, C, B, 1 cm, x, A

10 **P** Here are the lengths of sides of triangles.
Which of these triangles are right-angled triangles?
a 2 cm, 3 cm, 5 cm
b 6 cm, 8 cm, 10 cm
c 10 m, 24 m, 26 m

12.3 Trigonometry: the sine ratio 1

1 Copy these right-angled triangles.
For each triangle, label the **hyp**otenuse 'hyp' and the side **opp**osite the angle θ 'opp'.
The first has been started for you.

a
hyp, θ

b
θ

c
θ

2 a **R** Draw these triangles accurately using a ruler and protractor.

A
4 cm, 30°

B
12 cm, 30°

C
20 cm, 30°

b Label the hypotenuse (hyp) and opposite side (opp).
c Measure the opposite side.
d i Write the fraction $\frac{\text{opposite}}{\text{hypotenuse}}$ for each triangle and convert it to a decimal.
ii What do you notice?
iii What do you think the fraction will be for any right-angled triangle with an angle of 30°?

3 Use your calculator to find, correct to 3 d.p.
a sin 45°   b sin 81°   c sin 18.3°

4 Write sin θ as a fraction for each triangle.

a
11 cm, 9 cm, θ

b
9 m, 15 m, θ

c
θ, 12 mm, 17 mm

5 Find the value of x in each triangle. Give your answers correct to 1 d.p.

Example
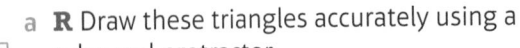

a
25°, 6 cm, x

b
72°, x, 9 cm

c
18°, x, 3.2 km

d
56°, x, 5.4 m

e
4.6 mm, 27°, x

6 A fishing line is cast from the top of a 3 m wall. The line makes an angle of 55° with the water. How long is the fishing line?

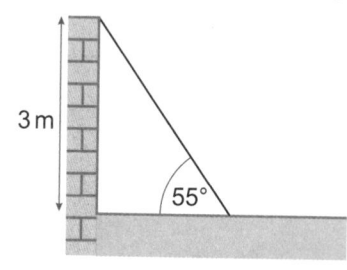

7

> **Exam-style question**
>
>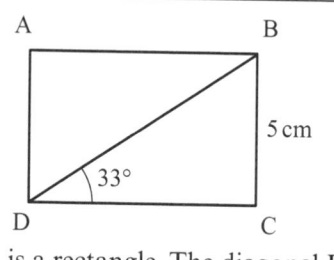
>
> ABCD is a rectangle. The diagonal BD is drawn and makes an angle of 33° with the line CD.
>
> Given that BC = 5 cm work out the length of BD. **(3 marks)**

8 **R** A wheelchair ramp is set to an angle of 6° with the horizontal.
The ramp must go up a step 0.5 m high.
The ramp is 6 m long. Is it long enough?

> **Q8 hint** Draw a diagram to help visualise the question.

12.4 Trigonometry: the sine ratio 2

1 Find angle θ. Each one is a multiple of 10°.
  a $\sin \theta = 0.766\ 044\ 4431$
  b $\sin \theta = 0.984\ 807\ 753$
  c $\sin \theta = 0.342\ 020\ 1433$

> **Q1 hint** Use the $\boxed{sin}$ key on your calculator. You only need to try angles such as 10°, 20°, 30°, …

2 Find angle θ. Each one is a multiple of 5°.
  a $\sin \theta = 0.707\ 106\ 7812$
  b $\sin \theta = 0.996\ 194\ 6981$
  c $\sin \theta = 0.087\ 155\ 7427$

3 Use $\sin^{-1}$ on your calculator to check your answers to **Q1** and **Q2**.
The first has been started for you.
  a $\sin \theta = 0.7660444431$
     $\theta = \sin^{-1}(0.7660444431)$
     $\theta = \square$

4 Use $\sin^{-1}$ on your calculator to find the value of θ correct to 0.1°.
  a $\sin \theta = 0.678$
  b $\sin \theta = 0.9214$
  c $\sin \theta = 0.7777$

5 Use your calculator to find the value of θ correct to 0.1°.
  a $\sin \theta = \frac{5}{6}$ b $\sin \theta = \frac{2}{15}$ c $\sin \theta = \frac{3.2}{19.8}$

6 Copy and complete these diagrams. The first one has been done for you.
  a $\sin 45°$

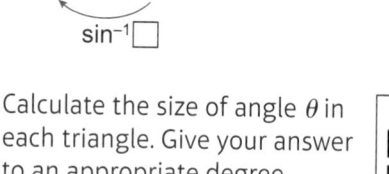

  b $\sin 65°$

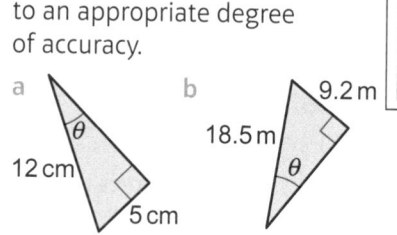

  c $\sin \square$

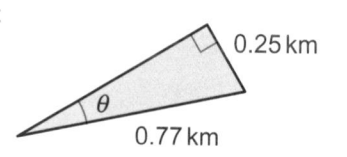

7 Calculate the size of angle θ in each triangle. Give your answer to an appropriate degree of accuracy.

Example

a 12 cm, 5 cm, θ

b 18.5 m, 9.2 m, θ

c 0.25 km, 0.77 km, θ

8 **P** An 18 m ramp rises 1.8 m.
What angle does the ramp make with
the horizontal?

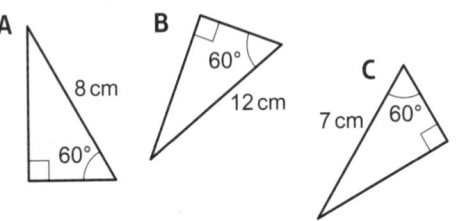

18 m

1.8 m

x

9

C

9 cm

5 cm

x

A   B

ABC is a right-angled triangle.

BC = 5 cm

AC = 9 cm

Calculate the size of the angle marked x.

Give your answer correct to the nearest
1 decimal place.   **(3 marks)**

12.5 Trigonometry: the cosine ratio

1 **a** **R** Draw these triangles accurately using a
ruler and protractor.

A   **B**

60°

8 cm   12 cm

C

60°   7 cm 60°

b Label the hypotenuse (hyp) and adjacent
side (adj).

c Measure the adjacent side.

d **i** Write the fraction $\dfrac{\text{adjacent}}{\text{hypotenuse}}$ for each
triangle and convert it to a decimal.

ii What do you notice?

2 Use your calculator to find, correct to 3 d.p.

a cos 30°

b cos 1°

c cos 19.5°

d cos 87°

e cos 19.5°

f cos 75°

3 Write cos θ as a fraction for each triangle.

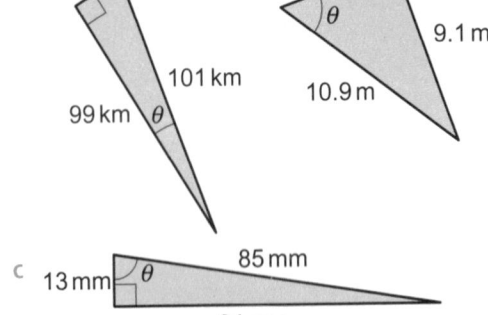

a 20 km

b 6 m

θ

9.1 m

101 km

10.9 m

99 km θ

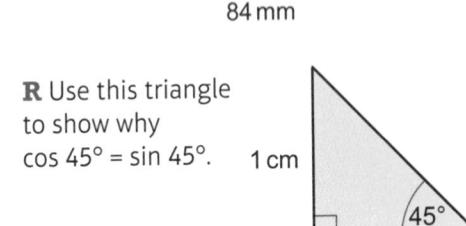

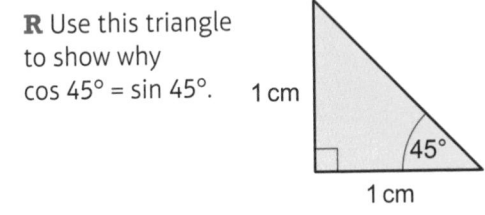

85 mm

c 13 mm θ

84 mm

4 **R** Use this triangle
to show why
cos 45° = sin 45°.

1 cm

45°

1 cm

5 Find the value of x in each
triangle. Give your answers
correct to 1 d.p.

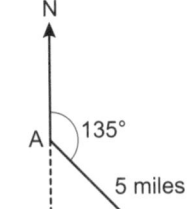

a 12

35°

x

b x

64°

45

c x

45°

15

6 **P** A man walks on a bearing
of 135° from point A for
5 miles.
How far south has
he travelled?

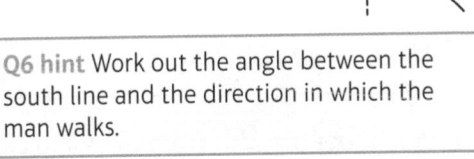

N

A 135°

5 miles

Q6 hint Work out the angle between the
south line and the direction in which the
man walks.

7 Use cos⁻¹ on your calculator to find the value
of θ correct to 0.1°.

a cos θ = 0.75

b cos θ = 0.921

c cos θ = 0.12

8 Use your calculator to find the value of θ correct to $0.1°$.

a $\cos \theta = \dfrac{3}{7}$

b $\cos \theta = \dfrac{2}{19}$

c $\cos \theta = \dfrac{84.5}{96.1}$

9 Copy and complete these diagrams.

a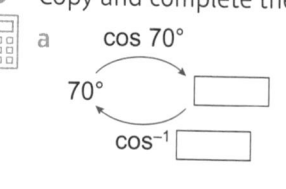
$\cos 70°$
$70°$
$\cos^{-1}$

b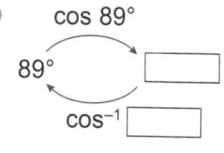
$\cos 89°$
$89°$
$\cos^{-1}$

c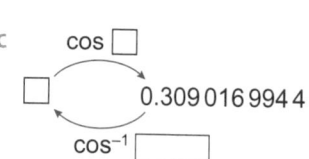
$\cos \square$
$\square$
$0.309\,016\,994\,4$
$\cos^{-1}$

10 Calculate the size of angle θ in each of these triangles. Give your answers to 1 d.p.

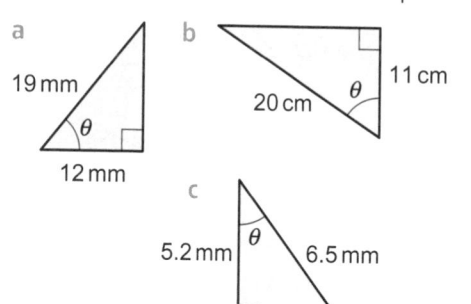

a
$19\,\text{mm}$
θ
$12\,\text{mm}$

b
$11\,\text{cm}$
θ
$20\,\text{cm}$

c
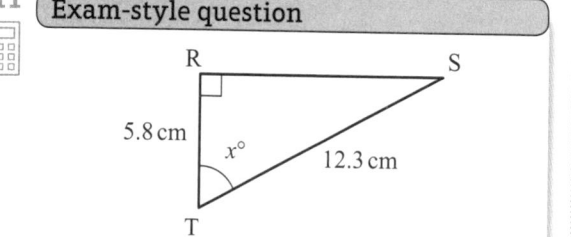
$5.2\,\text{mm}$
θ
$6.5\,\text{mm}$

11

Exam-style question

R
S
$5.8\,\text{cm}$
$x°$
$12.3\,\text{cm}$
T

RST is a right-angled triangle.

ST = 12.3 cm

RT = 5.8 cm

Calculate the size of the angle marked $x°$.

Give your answer to 1 decimal place.

(3 marks)

1 a **R** Draw these triangles accurately using a ruler and protractor.

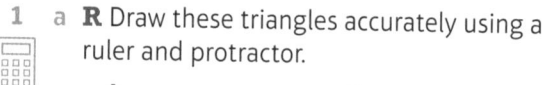

A
$20\,\text{cm}$
$45°$
x

B
x
$45°$
$12\,\text{cm}$

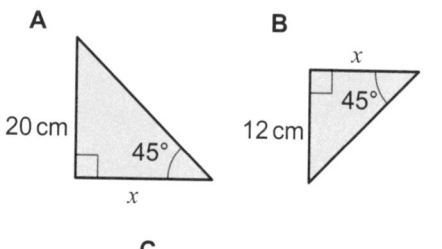

C
$45°$ $5.4\,\text{cm}$
x

b Label the opposite (opp) and adjacent side (adj).

c Measure the side marked x.

d i Write the fraction $\dfrac{\text{opposite}}{\text{adjacent}}$ for each triangle and convert it to a decimal.

  ii What do you notice?

2 Use your calculator to find, correct to 3 d.p. where necessary

a tan 30°

b tan 46°

c tan 73°

d tan 12°

e tan 85°

f tan 19.2°

3 Write tan θ as a fraction for each triangle.

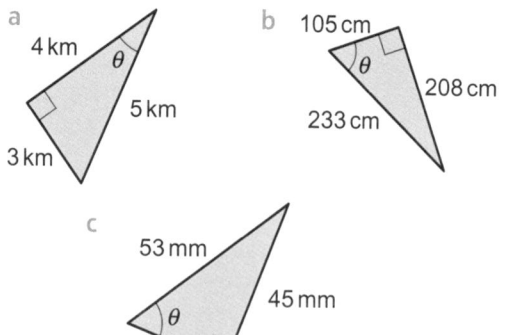

a
$4\,\text{km}$
θ
$5\,\text{km}$
$3\,\text{km}$

b
$105\,\text{cm}$
θ
$208\,\text{cm}$
$233\,\text{cm}$

c
$53\,\text{mm}$
$45\,\text{mm}$
θ
$28\,\text{mm}$

4 Find the value of x in each triangle. Give your answers correct to 1 d.p.

Example

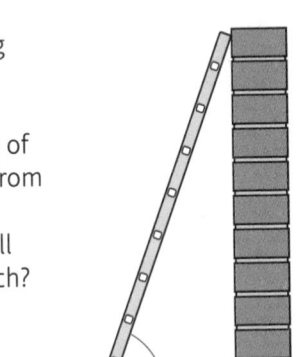

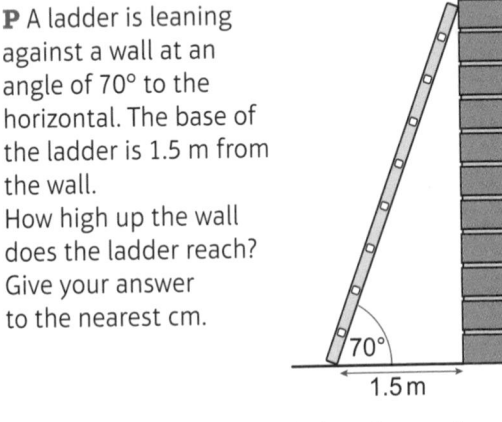

5 **P** A ladder is leaning against a wall at an angle of 70° to the horizontal. The base of the ladder is 1.5 m from the wall.
How high up the wall does the ladder reach?
Give your answer to the nearest cm.

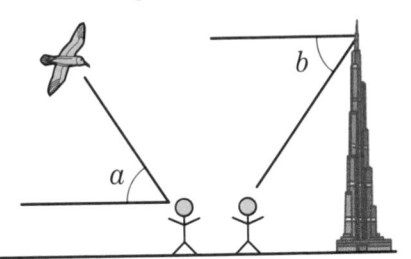
1.5 m

6 Which angle is an angle of elevation, and which is an angle of depression?

7 **P** The angle of elevation of the top of a set of rugby posts, C, from the kicker's kicking tee, B, is 37°. The kicking tee is 22 metres from the base of the posts.

C

B   A

a Copy the diagram and label it with this information.

b How tall are the posts?
Give your answer correct to the nearest cm.

8 **P** From the top of a lighthouse, Mr Amesbury sees a surfer. The angle of depression from Mr Amesbury to the surfer is 40°. The lighthouse is 60 m above sea level.
How far from the base of the lighthouse is the surfer?

9 Use $\tan^{-1}$ on your calculator to find the value of θ correct to 0.1°.

a $\tan \theta = 0.91$ b $\tan \theta = 1.49$

c $\tan \theta = \dfrac{11}{5}$ d $\tan \theta = \dfrac{0.9}{0.77}$

10 Copy and complete these diagrams.

a $\tan 55°$

55° ⟶ ☐

$\tan^{-1}$ ☐

b $\tan \square °$

65° ⟶ ☐

$\tan^{-1}$ ☐

c $\tan \square$

☐ ⟶ 1.732050808

$\tan^{-1}$ ☐

11 Calculate the size of angle θ in each of these triangles.

a

367 cm
θ
23 cm

b

θ
12.5 m
13.4 m

c

θ
$\sqrt{7}$
$\sqrt{2}$

12 **Exam-style question**

ABC is a right-angled triangle.
AB = 19.5 cm
BC = 24 cm
Calculate the size of the angle marked y.
Give your answer to 1 decimal place.

C

24 cm

B   19.5 cm   A
y

(3 marks)

12.7 Finding lengths and angles using trigonometry

1 Calculate the value of x in each triangle. Give your answers correct to 3 s.f.

Example

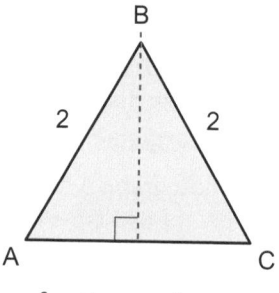

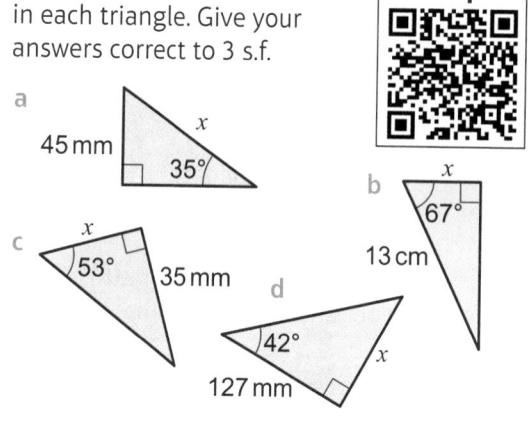

a

45 mm x 35°

c x 53°   35 mm

b x 67°   13 cm

d   42° x 127 mm

2 Calculate the size of angle x in each of these triangles.

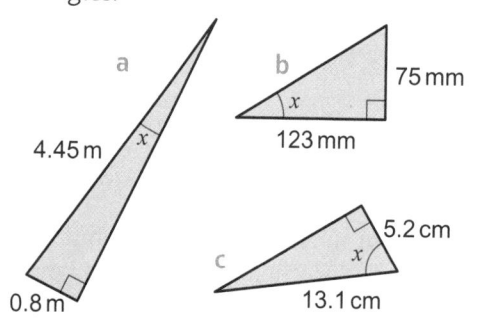

a   4.45 m x 0.8 m

b   75 mm x 123 mm

c   5.2 cm x 13.1 cm

3

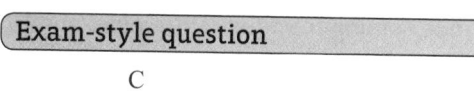
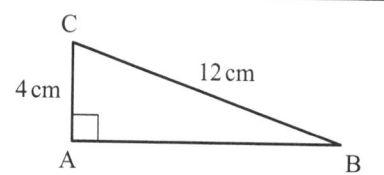

Exam-style question

C   4 cm   12 cm   A   B

ABC is a right-angled triangle.

AC = 4 cm

BC = 12 cm

Work out the size of angle ACB.

Give your answer correct to 1 decimal place. **(3 marks)**

4 **P** The Eiffel Tower is 301 m tall. A man stands 40 m from the base of the tower and looks at the top.

Find the angle of elevation of his sight line.

5 **P** A diver spots a ship wreck from the surface of the water. The angle of depression is 47°. The water is 50 m deep.

How far will he have to swim on the surface until he is directly over the wreck?

6 **R** Triangle ABC is an equilateral triangle with side length 2.

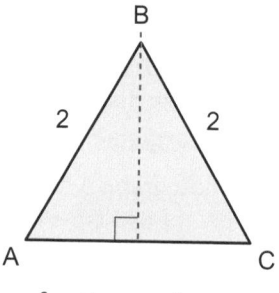

Work out

a   i   sin 30°

    ii   cos 30°

    iii   tan 30°

b   i   sin 60°

    ii   cos 60°

    iii   tan 60°

Express your answers as fractions, using surds when necessary.

7 **R** Triangle PQR is a right-angled isosceles triangle. PQ and PR have length 1.

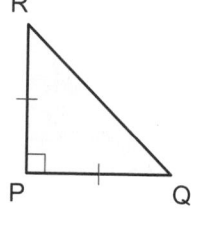

Work out

a   sin 45°

b   cos 45°

c   tan 45°

Express your answers as fractions, using surds when necessary.

8 Using your answers to **Q6** and **Q7**, find the lengths of the sides marked with letters.

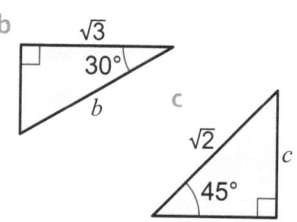
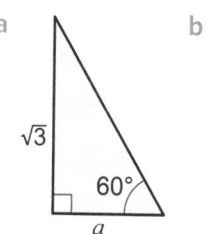

a $\sqrt{3}$ 60° a

b $\sqrt{3}$ 30° b

c $\sqrt{2}$ 45° c

9 **R** Triangle ACD is equilateral with sides of length 2 cm. Work out the length of BD.

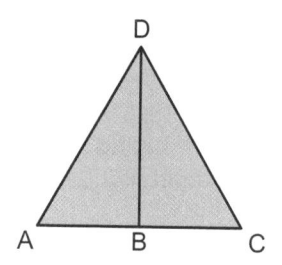

10 A vertical tent pole is 2.2 m high. The guy rope for the pole is tied to a peg 2.2 m from the bottom of the pole.

What is the angle of elevation from the tent peg to the top of the pole?

12 Problem-solving

Solve problems using these strategies where appropriate:

Example

- **Use pictures**
- **Use smaller numbers**
- **Use bar models**
- **Use x for the unknown**
- **Use flow diagrams**
- **Use formulae.**

1 Photograph A has a length of 8 cm and a height of 6 cm. Photograph B has a length of 15 cm and a height of 11 cm. Photograph C is an enlarged copy of photograph A and has a height of 18 cm.

  a How much longer is photograph C than photograph B?

  b What is the scale factor of enlargement from A to C?

2 There are 168 students in Year 9. The students can choose to study either French or German. The ratio for choice of French : German is 5 : 3. Class sizes can be a maximum of 30 students. How many classes are needed for the students studying French?

3 The lengths of the sides of four triangles are given below in cm to 1 decimal place. Which of these triangles are right-angled triangles?

  A 8.5, 11.0, 13.9

  B 74.0, 120.0, 135.0

  C 22.4, 15.5, 27.2

  D 14.7, 7.2, 15.5

> **Q3 hint** Arrange the information in a table to see it more clearly.

4 **R** Jabina's allotment is in the shape of a right-angled triangle. She is putting up a fence on all three sides of the allotment. The two short sides are 14.4 m and 8.5 m in length. If fencing is sold in whole numbers of metres only, what length of fencing should Jabina buy to have the smallest amount left over?

5 **R** Johan needs to fix a window at the front of a hotel. The window is located about 7 metres from the ground. To avoid the hotel door frame, Johan must position the bottom of his ladder 2 metres from the front of the hotel. Is an 8 metre ladder long enough? How do you know?

6 Yan draws two line segments. The first line segment starts at (6, 2) and ends at (2, 5), and the second line segment ends at (7, 4) and starts at (3, 5). Find the coordinates of the midpoint of each line segment.

7 **R** The diagram shows 3 paths on a nature walk.

Eagle path — Fox path — Squirrel path

  a How do you know without calculating which path is the longest?

  b Eagle path is about 2.4 km long and Fox path is about 7.5 km long. About how long is Squirrel path? Give your answer to 1 decimal place.

8
> **Exam-style question**
>
> A —— 22.5 cm —— B
> 12.0 cm
> C
>
> ABC is a right-angled triangle.
>
> AB = 22.5 cm
>
> BC = 12.0 cm
>
> Work out the length AC. Give your answer to 1 decimal place. **(3 marks)**

9 Paulina is putting up her tent. She has a guy rope coming from the centre pole. The pole is 180 cm tall and the rope is 360 cm long.

  a What angle from the ground will the rope be?

  b How far from the pole will the rope be pegged out to? Give your answer to to 1 d.p.

10 Satoshi is building a ramp. He has a 75 cm plank that goes up to the step. The plank ends at a distance of 60 cm from the bottom of the step.

  a What is the angle of the ramp? Give your answer to 1 d.p.

  b How high is the top of ramp above the ground?

13 PROBABILITY

13.1 Calculating probability

1 **R** A fair 6-sided dice is rolled.
   a Are the events 'an even number' and 'an odd number' equally likely?
   b Which is more likely, a number greater than 3 or a number less than 2?
   Explain your answers.

2 Look at this fair 5-sided spinner.

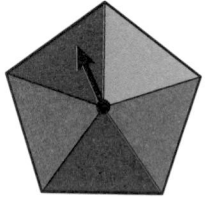

   a Write the probability of the spinner landing on blue.
   b Write the probability of the spinner landing on either blue or red.
   c Write the probability of the spinner landing on yellow.
   d Which colour is half as likely as blue?

3 The letters from the word ATTENDANCE are written on cards and placed in a bag.
   Mal picks one card at random from the bag.
   Work out
   a P(D)
   b P(A)
   c P(E or N)

4 In a class there are 7 students with blue eyes, 8 students with brown eyes and 3 students with green eyes.
   A student is picked at random.
   What is the probability that the student
   a has blue eyes
   b has green eyes
   c does *not* have green eyes?

5 A bag contains coloured balls.
   The probability of picking a red ball is $\frac{3}{11}$.
   What is the probability of picking a ball that is *not* red?

6 **Exam-style question**
   There are 15 coloured balls in a bag.
   3 are red, 5 are blue and 7 are black.
   Asha takes a ball from the bag at random.
   Write the probability that Asha
   a takes a red ball **(1 mark)**
   b does *not* take a black ball **(1 mark)**
   c takes a white ball. **(1 mark)**

7 **R** Donnie has a spinner with 12 different sections coloured red, blue and green.
   a P(red) = $\frac{6}{12}$ and P(green) = $\frac{2}{12}$.
   What is P(blue)?
   b Donnie changes all the sections, adding more red sections so that the probability of spinning blue is now $\frac{1}{4}$.
   How many red sections does he add?

8 A fair 6-sided dice is rolled. Are the following pairs of events mutually exclusive?
   a Rolling an odd number and rolling a multiple of 2
   b Rolling an odd number and rolling a prime number
   c Rolling an even number and rolling a multiple of 3

Q8a hint Can you roll a number that is both odd and a multiple of 2 at the same time?

9 A fair 12-sided dice numbered 1 to 12 is rolled.
   Are these pairs of events exhaustive?

   a Rolling an odd number and rolling an even number
   b Rolling a prime number and rolling a factor of 12
   c Rolling a multiple of 2 and rolling a number less than 11
   d Work out
   i P(prime number)
   ii P(factor of 6)
   iii P(number less than 12).

10 The probability that it will rain tomorrow is $\frac{1}{12}$.
   The probability that it will snow tomorrow is $\frac{1}{3}$.
   Work out the probability that it will not rain or snow tomorrow.

11 A 4-sided dice has the numbers 1 to 4 on its faces. The table shows the probabilities of rolling 1, 2 and 4.

Outcome	1	2	3	4
Probability	0.5	0.1		0.2

a Work out P(3).

b Write P(prime number) as

i a fraction

ii a percentage.

c Harrison rolls the dice 200 times. Predict the number of times he will roll 2.

12 P A 2-character password is made up of a vowel followed by a single digit.

a How many possible combinations are there?

b What is the probability of someone getting the password right first time by chance?

c It is known that the password does not contain an O or a U. What is the probability of someone getting the password right first time by chance?

13.2 Two events

1 This spinner is spun twice.

Example

a Copy and complete the sample space diagram to show the possible outcomes.

	2nd spin	
	Red	**Blue**
1st spin **Red**	R, R	
1st spin **Blue**		

b How many possible outcomes are there?

c Write the probability of getting red twice.

d Write the probability of getting one red and one blue.

e Write the probability of getting at least one blue.

2 R A fair dice is rolled and a coin is flipped.

 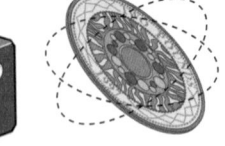

a Copy and complete the sample space diagram showing the possible outcomes.

	Head	**Tail**
1	1, H	1, T
2	2, H	
3		
4		
5		
6		

Work out

b P(1, H)

c P(odd number, H)

d P(number less than 6, T)

e Jayshuk says, 'The probability of getting an even number and a tail is $\frac{1}{2}$ since you will get an even number $\frac{1}{2}$ of the time and get a tail $\frac{1}{2}$ of the time.'

Explain why Jayshuk is wrong.

> **Q2e hint** Show any working then write 'Jayshuk is wrong because …'

3 These two spinners are spun.

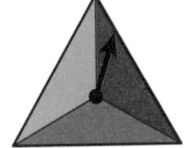

a Draw a sample space diagram to show all the possible outcomes.

b Find the probability that both spinners land on the same colour.

c Predict the number of times both spinners will land on blue in 60 spins.

Exam-style question

A company makes
T-shirts in five different
colours: yellow, red,
blue, green and white.

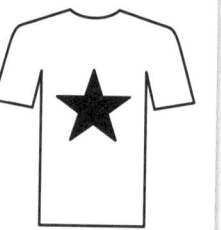

They let you choose
between two different
logos: a star or a tree.

One possible T-shirt is shown.

a List all the possible T-shirts you
could order.
The first one is done for you.
(white, star) **(2 marks)**

b A T-shirt is picked at random.
Work out the probability that the T-shirt
is blue with a tree on the front. **(1 mark)**

5 A necklace is made using
two beads.
There is a choice of gold,
silver and emerald beads.

a Draw a sample space diagram showing all
the options.

b How many possible combinations
are there?

c A necklace is picked at random. Write the
probability that it has
  i only gold beads
  ii a silver bead and an emerald bead
  iii at least one emerald bead.

6 Two fair dice numbered 1–6 are rolled and
the product of the scores is found.

a Copy and complete the sample space
diagram showing the possible total scores.

Dice 1	1	2	3	4	5	6
1						
2	2	4				
3			9	12		
4						
5				20		
6						

Dice 2

b Find the probability of scoring
  i a total of 6   **ii** more than 20
  iii an even number   **iv** a multiple of 3.

c Which is more likely, scoring a multiple of 3
or a multiple of 5?

7 **P** Sam puts three £1 coins and a 50p coin
in one pocket. He puts a £1 coin and two
10p coins in another pocket.
He takes a coin from each pocket at random.
What is the probability of him getting
more than £1?

8 **P** In a game, a player rolls a coin onto
a board.
The board has the values of £0.01,
£1 and £2 on it.
The probability of landing on each square is
equally likely.
If the coin lands heads up on a square,
the player wins the value shown in
the square.

£1	£0.01	£2
£2	£1	£0.01
£0.01	£2	£1
£1	£0.01	£2

The game costs £1 to play. Kali thinks that a
player is more likely to lose money than win.
Is she correct?

9 These two spinners are spun and the scores
are added.

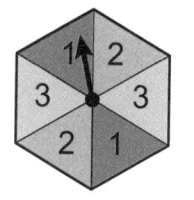 

Work out the probability that the total score
is more than 4.

13.3 Experimental probability

1 Amery spun this spinner 20 times.

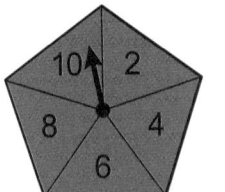

a Copy and complete the frequency table to record his outcomes.

4, 6, 2, 2, 6, 4, 6, 6, 8, 2,
10, 2, 6, 6, 4, 2, 6, 8, 8, 6

Outcome	2	4	6	8	10
Frequency					

b What does the frequency tell you? Write a definition in your own words.

2 Ethan drops a piece of toast 50 times.
He records whether it lands 'butter side down' or butter side up'.
He records his results in a table.

	Down	Up
Frequency	27	23

The relative frequency of 'butter side down' is $\frac{27}{50}$.
Write the relative frequency of 'butter side up'.

3 Sansa rolls a 6-sided dice.
She records her results in a table.

	1	2	3	4	5	6
Frequency	17	19	11	16	15	22

a How many trials did Sansa do?
b What is the relative frequency of her scoring 6?

4 Inu and James are recording the colours of cars that pass by.
Their results are recorded in the table.

	Red	Blue	Black	White	Grey	Green
Inu	12	14	20	21	32	1
James	15	29	23	29	41	3

a How many cars did Inu record?
b Write the estimated probability of a black car from Inu's results.
c How many cars did James record?
d Write the estimated probability of a black car from James's results.
e Altogether Inu and James recorded 240 cars.
What is the estimated probability of a black car from their combined results?
f Which is the best estimate for P(black)? Explain.

5 Vinny thinks a coin is biased. He flips it 20 times and it lands heads up 13 times.
He thinks this shows that the coin is biased.
Do you agree? How could he get a more accurate measure of probability?

6 Martin spins this spinner and records the number spun each time.

	1	2	3
Frequency	13	9	8

a What is the experimental probability of Martin scoring a 1?
b What is the theoretical probability of Martin scoring a 1 if the spinner is fair?

7 A game is played where a coin is flipped and an ordinary 6-sided dice is rolled.
A player wins if they get a head and an even number.
The table shows the results of 40 games.

	Win	Lose
Frequency	8	32

a What is the experimental probability of winning?
b What is the theoretical probability of winning?
c How many wins would you predict in 40 turns?
d Do you think either the dice or the coin is biased? Explain your answer.

8 **P** At a town fayre a game is played where a coin is flicked onto a board.

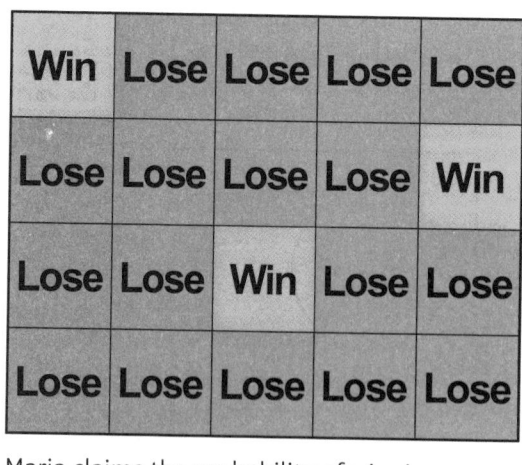

Win	Lose	Lose	Lose	Lose
Lose	Lose	Lose	Lose	Win
Lose	Lose	Win	Lose	Lose
Lose	Lose	Lose	Lose	Lose

Maria claims the probability of winning on any go is $\frac{1}{10}$. She charges 50p per go and gives £1 for each winning throw.

a Maria expects 100 people to play. If her claim is true, how much money will she make?

b The table shows the actual results.

	Win	Lose
Frequency	17	83

Estimate the probability of winning from this information.

9

a Sarah spins a biased spinner numbered 1–4 50 times. The table shows her results. She spins the spinner once more. Find an estimate for the probability that she will get a 4. **(1 mark)**

Score	Frequency
1	22
2	9
3	9
4	10

b Mark has a biased spinner numbered 1–2. The probability that he will spin a 1 is 0.4. Mark is going to spin the spinner 200 times. Work out an estimate for the number of times he will spin a 1. **(2 marks)**

Exam hint
Show your working. If you calculate your answer incorrectly, you may still get marks for method.

10 The table shows some information about parcels posted at a particular post office.

	UK	Overseas	Total
First class	75	18	93
Second class	91	4	95
Total	166	22	188

a Work out the probability that a parcel picked at random is being posted
  i first class
  ii second class going overseas.
b A second class parcel is picked at random. What is the probability that it is being posted overseas?

11 Lisa records the type and colour of vehicles passing her house on Saturday afternoon. The table shows her results.

	Red	Black	White	Other
Bus	16	1	5	3
Van	2	0	13	3
Car	4	10	8	25

a Estimate the probability that the next car to go past Lisa's house is
  i a white van   ii a red bus.
b Which is more likely to pass her house, a red car or a white bus?
c Which type of vehicle is most likely to be white?
d Which type of vehicle is most likely to be red?
e What is the most likely car colour?

13.4 Venn diagrams

1 Set A = {even numbers less than 10}
Set B = {prime numbers less than 10}
a List the numbers in each set:
  A = {2, 4, …}, B = {2, 3, …}
b Write 'true' or 'false' for each statement.
  i 8 ∈ A   ii 9 ∈ B   iii 12 ∈ B
c Which number is in both sets?
d Which numbers are in set A only?
e Which numbers are in either A or B (or both)?
f Which numbers less than 10 are not in A or B?

2 The Venn diagram shows 2 sets, A and B, and ℰ, the set of all numbers being considered.

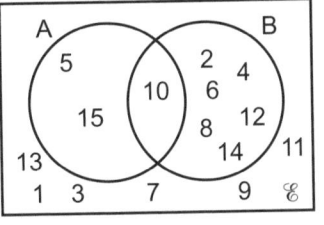

a Copy and complete these sets.
A = {5, ☐, ☐}
B = {2, 4, ☐, ☐, ☐, ☐, ☐}
ℰ = {1, 2, 3, …}

b Match each set to its description.

| A | {integers 1 to 15} | B | {multiples of 5 up to 15} | ℰ | {multiples of 2 up to 15} |

> **Q2a hint** ℰ includes all the numbers in A and B too.

3 Copy and complete the Venn diagram for these sets.
ℰ = {integers from 1 to 20}
X = {even numbers from 1 to 20}
Y = {odd numbers from 1 to 20}

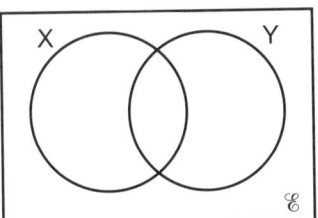

4 For the Venn diagram in **Q2**, copy and complete these sets.
a A ∩ B = {☐}
b A ∪ B = {2, 4, …}
c A′ = {1, …}
d B′ = {1, …}
e A′ ∩ B = {…}

5 For the Venn diagram you drew in **Q3**, write these sets.
a X ∩ Y
b X ∪ Y
c X′
d Y′
e X′ ∩ Y
f X ∪ Y′

6 The owner of a bed shop records the number of customers and the numbers of mattresses and beds sold in one week. The Venn diagram shows the results.

Example

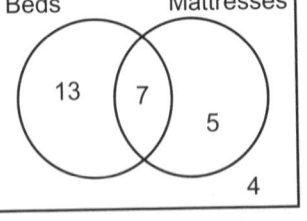

a How many beds were sold?
b How many customers bought a bed, a mattress or both?
c What is the probability that a customer picked at random bought a bed and a mattress?
d What is the probability that a customer picked at random bought only a mattress?
e A shop assistant says, 'The Venn diagram shows that five mattresses were sold.' Explain why he is wrong.

7 The manager of a coffee shop records customers' purchases before 11 am. Of the 100 customers surveyed, 37 buy only a drink and 23 buy a drink and a snack. All customers buy something.
a Draw a Venn diagram to show the manager's findings.
b What is the probability that a customer picked at random buys only a drink?
c What is the probability that a customer picked at random buys a snack?

8 Exam-style question

B is the set of students studying biology.
C is the set of students studying chemistry.
The Venn diagram shows the number of students in each set.

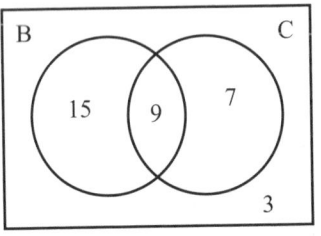

a Work out P(B ∩ C). (2 marks)
b Work out P(B′ ∪ C′). (2 marks)

13.5 Tree diagrams

1 In a pack of yogurts, 2 are strawberry and 4 are raspberry.

 a What is the probability of picking a raspberry yogurt?

 Angelica picks a yogurt at random. It is raspberry and she eats it.

 b How many yogurts are left in the pack?

 c How many raspberry yogurts are left in the pack?

 She picks another yogurt at random.

 d What is the probability of picking a raspberry yogurt this time?

2 Which of these pairs of events are independent?

 A Taking a chocolate from a box, eating it, and then taking another chocolate from the box.

 B Getting a head on the first flip of a coin and then getting a tail on the second flip.

 C Rolling a prime number on a dice and then rolling an odd number.

 D Rolling a 6 on a dice and then rolling another 6.

3 Livia has a bag containing 3 red marbles and 4 blue marbles.

 She picks a marble at random, records the colour, puts the marble back in the bag and then picks another marble at random. The frequency tree shows the possible outcomes.

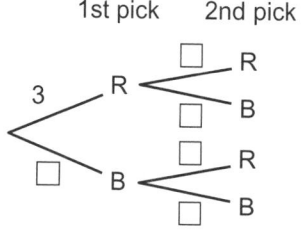

 a Copy the frequency tree.

 b 3 is on the first branch as there are 3 red marbles.
 How many blue marbles are there?
 Use your answer to complete the branches for the 1st pick on the frequency tree.

 c Livia puts the marble back in the bag.
 How many red and blue marbles are in the bag when she picks another one?
 Complete the branches for the 2nd pick by filling in the rest of the numbers.

4 A bag contains 10 apples. 4 are rotten and 6 are fresh.

 Henry picks an apple from the bag, replaces it and then takes a second apple.

 a Copy and complete this frequency tree.

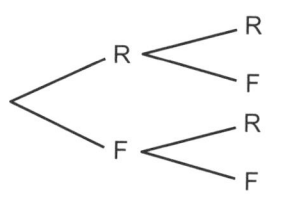

 b How many ways are there of picking two fresh apples?

> **Q4 hint**
>
>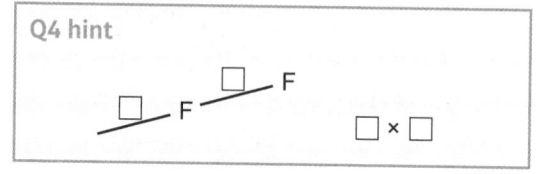

5 A box of sweets contains 10 toffees and 5 chocolates.

 Yvonne picks a sweet at random, puts it back and picks another sweet.

 Draw a frequency tree to show this.

6 The tree diagram shows the probabilities that Aman will roll a 6 during two rolls of a dice.

Example

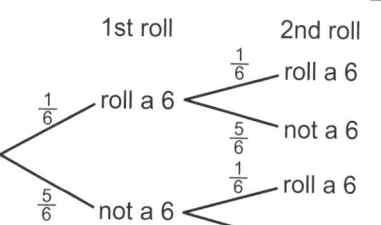

 a What is the probability he rolls a 6 on the first roll?

 b What is the probability that he rolls two 6s?

 c What is the probability that he doesn't roll a 6 in either roll?

 d What is the probability that he rolls only one 6? Explain your answer.

7 Abbot's phone signal is unreliable. The probability he has a signal on any day is $\frac{4}{5}$.

a Write the probability that he doesn't have a signal.

b Copy and compete the tree diagram.

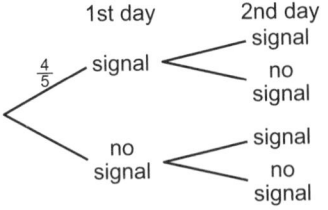

1st day   2nd day

$\frac{4}{5}$ signal — signal / no signal

no signal — signal / no signal

c Work out the probability that he has a signal for two days in a row.

d Work out the probability that he has no signal on the first day but a signal on the second day.

8 A box of chocolates contains 3 dark chocolates and 12 milk chocolates.
Mr Murphy picks a chocolate at random, notes the type and puts it back the box. He then picks another chocolate.

a Copy and complete the tree diagram.

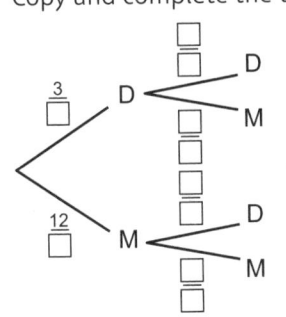

$\frac{3}{\square}$ D — D / M

$\frac{12}{\square}$ M — D / M

b What is the probability that Mr Murphy picks one chocolate of each type?

9 ⬚ **Exam-style question**

Ganit flips two fair coins.

first coin   second coin

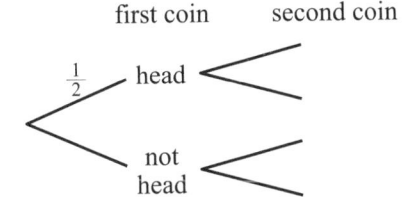

$\frac{1}{2}$ head

not head

a Copy and complete the probability tree diagram to show the outcomes. Label clearly the branches of the probability tree diagram. **(2 marks)**

b Calculate the probability that Ganit gets a head on both coins. **(1 mark)**

c Calculate the probability that Ganit gets at least one head. **(2 marks)**

13.6 More tree diagrams

1 Are these pairs of events dependent or independent?

a Rolling a 3 on a dice and getting a tail when flipping a coin

b Picking a dark chocolate at random from a bag of sweets, eating it, then picking another dark chocolate

c Flipping a coin and getting a head, then flipping the coin again and getting another head

d Spending too much and getting into debt

2 Charlie's sock drawer contains 8 red socks and 5 blue socks.
The frequency tree shows what happens when he picks a sock at random, puts it on and then picks another sock.
Copy and complete his frequency tree.

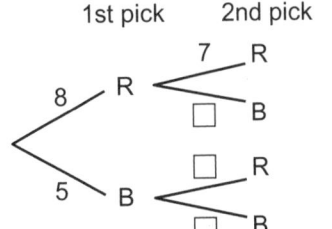

1st pick   2nd pick

8 R — 7 R / □ B

5 B — □ R / □ B

3 A bag of sweets contains 12 strawberry sherbets and 8 lemon sherbets.
Julie picks a sweet, eats it, and then picks another one.

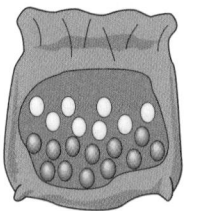

a Copy and complete the frequency tree to show this.

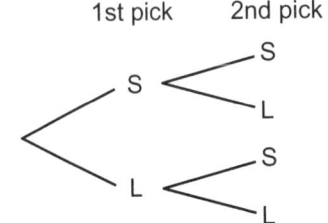

1st pick   2nd pick

S — S / L

L — S / L

b How many ways are there of picking two lemon sherbets?

4 Amy has a bag of coins containing seven £1 coins and three 20p coins.
She picks a coin at random, puts it in her pocket and then picks another coin.

Example

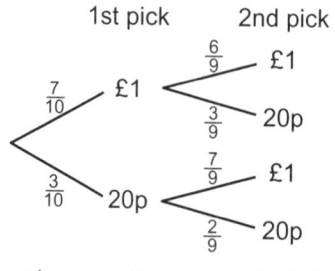

Use the tree diagram to find the probability that she picks

a two £1 coins

b one of each type of coin.

5 Qamar has these letter cards.

He picks two cards at random without replacing them.
He records whether the letter on the card is a vowel or a consonant.

a Copy and complete the tree diagram.

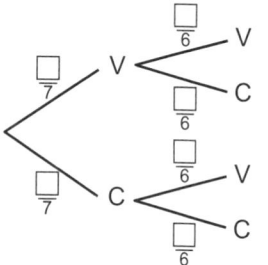

b Find the probability that he picks

i 2 vowel cards

ii 1 consonant card and 1 vowel card.

Q5 hint The vowels are A, E, I, O and U.

6 **R** Danny takes two buses to get to school.
The probability that the first bus is red is 0.4.
If the first bus is red, the probability that the second bus is red is 0.5.
If the first bus is green, the probability that the second bus is green is 0.4.

a Copy and complete the tree diagram.

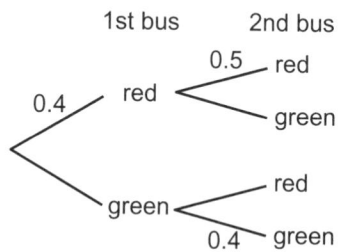

b What is the probability that both buses are red?

c What is the probability that neither bus is red?

d Write the probability that Danny gets at least one red bus.

7

Exam-style question

There are 20 biscuits in a tin.

12 of the biscuits are chocolate.

8 of the biscuits are plain.

Hameed takes at random two biscuits from the tin one after the other.

a Copy and complete the probability tree diagram.

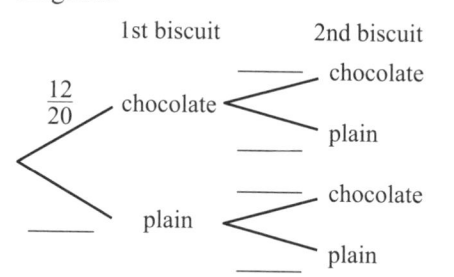

(2 marks)

b Work out the probability that Hameed takes two biscuits of the same type.
(3 marks)

13 Problem-solving

Solve problems using these strategies where appropriate:
- **Use pictures**
- **Use smaller numbers**
- **Use bar models**
- **Use x for the unknown**
- **Use a flow diagram**
- **Use more bar models**
- **Use formulae.**

1 The perimeter of each of these rectangles is 30 cm. How much longer is y than x?

5.5 cm [ ] [   ] 4.25 cm
     x y

2 Yohan has 1 hour until he needs to board his aeroplane. He spends $\frac{1}{5}$ of an hour shopping and $\frac{1}{2}$ hour eating lunch.
How long does Yohan now have left until he needs to board his aeroplane?

3 Omar says that when he rolls a fair 6-sided dice, the events of rolling a multiple of 3 and rolling an even number are mutually exclusive.
Is Omar correct? Use a Venn diagram to explain your answer.

4 Priya is playing a game. She drops a toy animal and wins points depending on which way it lands.
She gets most points if it lands on its nose and 0 points if it lands on its side.
She carries out an experiment to see if the probability of the event is related to the number of points given.
The table shows her results.

Outcome	Frequency
Side	48
Back	19
Feet	24
Nose	9

a How many times did Priya drop the toy animal?
b What is the experimental probability for each of the outcomes?
c Do you think the number of points relates to the probability of the event?
Explain your answer.

6 **R** A fair 6-sided number cube has the following properties.
- It has one number on each side.
- It has four different numbers.
- It has more sides with an odd number on.
- It has a probability of $\frac{1}{3}$ of rolling a 3.
What could be a possible set of numbers?

7 Susan has a drawer of coloured socks. She has two blue pairs, one green pair and three purple pairs. None of her socks have been put in pairs. She pulls out one sock at a time.
a What is the probability of pulling out a purple sock?
b What is the probability of not pulling out a green sock?
c Susan picks out a blue sock.
What is the probability of pulling out a second blue sock?

Q7 hint Can you draw a diagram to help solve this problem?

8 **R** The cross-section of a cuboid is 600 cm². The volume of the cuboid is 24 000 cm³.
The lengths of the sides of the cuboid are in the ratio $1 : 1.5 : 2$.
Work out the lengths of the sides.

9 **R** A teacher asks students in his class whether they like to sing or to play an instrument.
The Venn diagram shows the results.

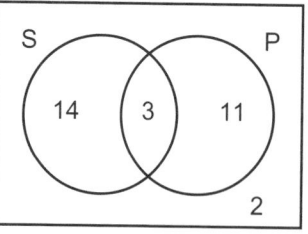

The teacher picks one person at random. Work out the probability that this person

a likes only singing

b likes to sing and to play an instrument

c likes to sing or play an instrument but not both

d likes neither singing nor playing an instrument.

Mike joins the class and the teacher asks him the same question. The probability that a person likes only to play an instrument is now $\frac{11}{31}$.

e What response could Mike have given?

10 **R** Freya and Lindsey play a game with two fair dice.
One dice has odd numbers 1–11 on it, the other dice has even numbers 2–12 on it.
They roll both dice at the same time, add the scores and record the total.
Freya wins a point if the total is less than 12. Lindsey wins a point if the total is 12 or more.

a What are all the possible totals?

b What is the probability of getting a total of 7?

c What is the probability of getting a total of an even number?

d Explain why the game isn't fair.

e Make up a game using the same dice which would be fair.

11 Simon is building a frame for the roof of his hen house.
One part of the frame will be a right-angled triangle. He knows that he wants the roof at an angle of 45° to the horizontal. The side opposite to this angle measures 55 cm.
What length of wood does Simon need for the longest side of the frame?
Round your answer to 1 d.p.

14 MULTIPLICATIVE REASONING

14.1 Percentages

1 The price of a washing machine is £450 after 20% VAT is added.
What was the price before VAT was added?

Example

2 A shop offers a 40% discount in a sale.
A jacket has a sale price of £27.
What was the original price?

3 A company made 4% more profit in 2014 than in 2013.
In 2014 it made a profit of £1331 200.
How much profit did the company make in 2013?

4

Exam-style question

The price of a bike is reduced by 20% in a sale.
The sale price of the bike is £480.
Work out the original price of the bike.
(3 marks)

5 **P** Mia buys a laptop that is reduced by 30%. She pays £455.

a What was its original price before the reduction?

b Libby pays £525 for a laptop that is reduced by 25%. Who saves more money?

6 **R** George earns £52 000 a year.

The first £10 000 of his earnings is free of tax.
He pays 20% income tax on the next £31 865.
He then pays 40% income tax on the rest.
George's employer deducts the tax monthly.
How much tax does George pay each month?

> **Q6 hint** Taxed pay = £52 000 − £10 000 = ☐
> Total tax for the year is
> 20% of £31 865 + 40% of (£52 000 − ☐ − ☐)
> To find the tax for a month, divide by ☐.

7 Rani invests £1800.

When her investment matures she receives £1917.
Copy and complete the calculation to find the percentage increase in her investment.
Actual change = £1917 − £1800 = £☐

$$\text{Percentage change} = \frac{\text{actual change}}{\text{original amount}} \times 100$$

$$= \frac{\square}{1800} \times 100 = \square\%$$

8 Josh invests £4400.

When his investment matures he receives £4294.40.
Calculate the percentage decrease in his investment.

9 Callum invests £2250.

When his investment matures he receives £2288.25.
Work out the percentage increase in his investment.

10 The table shows the prices a shop pays for some items (cost price) and the prices it sells them for (selling price).

Item	Cost price	Selling price	Actual profit	Percentage profit
shirt	£9	£15		
shoes	£21	£28		
jeans	£16	£24		
dress	£30	£36		

Copy and complete the table to work out the percentage profit on each item.

11 Victoria bought a car for £12 000.

Two years later she sold it for £9700.
Work out her percentage loss.
Give your answer to 1 d.p.

12 In 2003 the population of the USA was 288 998 781.

In 2013 it had increased to 315 079 109.
What was the percentage increase in the population?
Give your answer to 3 s.f.

13 **P** Jai buys 40 books for £3 each.

He sells $\frac{1}{2}$ of them for £7 each, $\frac{1}{4}$ of them for £5 each and the rest for £3.50 each.
What is his percentage profit?

14 **P / R** The table shows information about visitor numbers to a museum in 2013 and in 2014.

Year	Total number of visitors	Ratio of children to adults	Price of child ticket	Price of adult ticket
2013	14 760	1:3	£2.00	£7.20
2014	17 520	1:2	£2.50	£8.00

a Work out the percentage change in the total number of visitors from 2013 to 2014. Give your answer to 1 d.p.

b Does your answer to part **a** show a percentage increase or decrease?

c How many adults visited the museum in 2014?

d Work out the percentage change in the amount of money taken in ticket sales from 2013 to 2014. Give your answer to 1 d.p.

14.2 Growth and decay

1 Sean bought a car for £9000.

It lost 25% of its value in the first year.
It lost 15% of its value in the second year.

a What is the multiplier to find the value of the car at the end of the first year?

b What was the value of the car at the end of the first year?

c What is the multiplier to find the value of the car at the end of the second year?

d What was the value of the car at the end of the second year?

e What is the single decimal number that the original value of the car can be multiplied by to find its value at the end of the 2 years?

2 Raj has a job with an annual salary of £18 000.
At the end of the first year he is given a salary increase of 1.5%.
At the end of the second year he is given an increase of 2%.
  a Write the single number, as a decimal, that Raj's original salary can be multiplied by to find his salary at the end of the 2 years.
  b Work out Raj's salary at the end of 2 years.

3 **R** Georgia says, 'An increase of 10% followed by an increase of 18% is the same as an increase of 28%.'
Is Georgia correct? Explain your answer.

4 Jamie's manager says he can have either a 1% pay rise this year and then a 2.5% pay rise next year, or a 2% pay rise this year.
Which is the better offer? Explain.

5 **P** £4000 is invested for 2 years at 3% per annum compound interest.
Work out the total interest earned over the 2 years.

Example

6 £1700 is invested for 2 years at 2.5% per annum compound interest.
Work out the total amount in the account after 2 years.

7 £1200 is invested at 2% compound interest.
Copy and complete the table.

Year	Amount at start of year	Amount plus interest	Total amount at end of year
1	£1200.00	1200 × 1.02	£1224.00
2	£1224.00	1224 × 1.02 = 1200 × 1.02²	£1248.48
3	£1248.48	1248.48 × 1.02 =	
4	£1273.45		
5			

8 Work out the multiplier, as a single decimal number, that represents
  a an annual increase of 12% for 2 years
  b an annual decrease of 10% for 5 years
  c an increase of 7% followed by an increase of 3%
  d a decrease of 15% followed by a decrease of 8%
  e an increase of 11% followed by a decrease of 4%.

9 There are 1500 bacteria in a Petri dish.
The number of bacteria doubles every hour.
How many bacteria will be in the dish after 6 hours?

10 The level of activity of a radioactive source decreases by 10% per day.
The activity is 5000 counts per minute.
What will it be a week later?

11 The level of activity of a sample containing a radioactive isotope is 90 000 counts per minute.
The half-life is 3 days.
What will the count rate be after 12 days?

> **Q11 hint** The **half-life** is the time taken for the count rate to fall to half its starting value.

12 In 2014 a supermarket chain had 120 stores in the UK. The number of stores increases at a rate of 6% each year.
How many stores will it have in 2018?

13 **Exam-style question**

Liz makes an investment which will increase in value by 5% every year.
  a Liz says, 'After 10 years my investment will be worth 50% more than it is now.'
  Liz is wrong. Explain why. **(1 mark)**
Liz wants to work out the value of her investment after 2 years.
  b By what single decimal number should Liz multiply the value of the initial investment? **(2 marks)**

14.3 Compound measures

1 **R** Will has a basic pay rate of £8.80 per hour.
He is paid 1.5 times as much for every hour he works at the weekend and twice as much for every hour he works at night.
One week, Will works for 20 hours at his basic pay rate, 5 hours at the weekend and 8 hours at night.
How much does he get paid for the week?

2 Water leaks from a tank at a rate of 2 litres per hour.

  a Work out how much water leaks from the tank in
   i 30 minutes
   ii 45 minutes.

  Initially there are 70 litres of water in the tank.

  b Work out how long it takes for all the water to leak from the tank.

3

┌─────────────────────────────────────┐
Exam-style question

The diagram shows a fuel tank in the shape of a cuboid.

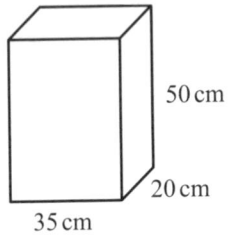

50 cm

20 cm

35 cm

The measurements of the cuboid are 35 cm by 20 cm by 50 cm.

  a Work out the volume of the tank.
   (1 mark)

Fuel is drained from the tank at a rate of 4 litres per minute.

1 litre = 1000 cm³

  b Work out the time it takes to empty the tank completely.
   Give your answer in minutes. **(2 marks)**
└─────────────────────────────────────┘

4 A car travels 500 km and uses 40 litres of petrol.

  a Work out the average rate of petrol usage. State the units with your answer.

  b Estimate the amount of petrol used when the car has travelled 75 km.

5 A block of lead has a mass of 1.5 kg and a volume of 130 cm³.
  What is its density in g/cm³?

6 A cubic metre of sand weighs 1500 kg.
  What is the density of the sand in g/cm³?

7 Copper has a density of 8.96 g/cm³.
  A piece of copper has a mass of 2.8 kg.
  Work out its volume.

Example

8 **R** 1 cm³ of tin has a mass of 7.37 g.
  1 cm³ of nickel has a mass of 8.91 g.

  a Write down the density of each metal.

  b Which metal is more dense?
   Explain your answer.

9 A piece of aluminium has volume 450 cm³ and density 2.7 g/cm³.
  Work out the mass of the aluminium.
  Give your answer in
  a grams
  b kilograms.

10 ┌─────────────────────────────────────┐
Exam-style question

The diagram shows a triangular prism.

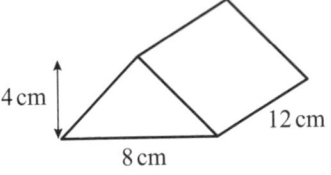

4 cm

12 cm

8 cm

The base of the triangle is 8 cm and its height is 4 cm.

The length of the prism is 12 cm.

  a Work out the volume of the prism.
   (3 marks)

The prism is made out of silver.
Silver has a density of 10.5 grams per cm³.

  b Work out the mass of the prism.
   Give your answer in kilograms. **(2 marks)**
└─────────────────────────────────────┘

11 a A force F of 50 N is applied to an area A of 1.4 m². Work out the pressure P in N/m².

  b A force applied to an area of 3.2 m² produces a pressure of 18 N/m².
   Work out the force in N.

┌─────────────────────────────────────┐
Q11b hint Substitute into the formula $P = \dfrac{F}{A}$
Rearrange to find F.
└─────────────────────────────────────┘

12 Copy and complete the table.

Force (N)	Area (m²)	Pressure (N/m²)
90	2.4	
60		9
	7.5	16

14.4 Distance, speed and time

1 Copy and complete the table.

Example

Distance (km)	Speed (km/h)	Time
105		1 h 30 min
	88	3 h 15 min
168	72	

2 Work out the average speed for these journeys.
  a A plane flies 750 miles in $1\frac{1}{2}$ hours.
  b A cyclist rides 27 miles in 2 hours 15 minutes.
  c A car travels 250 km in 3 hours 12 minutes.

3 Work out the distance travelled for these journeys.
  a A train travels for 2 hours 40 minutes at an average speed of 120 km/h.
  b A helicopter flies for 45 minutes at an average speed of 90 mph.
  c A man runs for 12 minutes at an average speed of 8.5 km/h.

4 Work out the time taken for these journeys.
  a A leopard walks 21 km at an average speed of 3.5 km/h.
  b A bee flies 0.25 km at an average speed of 5 m/s.
  c An ant walks 36 m at an average speed of 3 cm/s.

5 Convert these speeds from metres per second (m/s) to metres per hour (m/h).
  a 5 m/s
  b 16 m/s
  c 3 m/s

6 Hannah travels the first 80 km of a journey in 50 minutes. She then travels a further 30 km in half an hour.
  What is her average speed for the whole journey?

7 Peter walks 864 m from his house to school. The journey takes him 9 minutes.
  What is his average speed in m/s?

8
Exam-style question

On Saturday Laura cycles for 3 hours. Her average speed is 26 km/h.
  a How far does Laura cycle on Saturday? **(2 marks)**
On Sunday Laura cycles 55 miles.
5 miles ≈ 8 kilometres
  b On which day did Laura cycle further? **(3 marks)**

9 Convert these speeds from metres per second (m/s) to kilometres per hour (km/h).
  a 10 m/s
  b 3 m/s
  c 12 m/s

10 Convert these speeds from kilometres per hour (km/h) to metres per second (m/s).
  a 18 km/h
  b 90 km/h
  c 63 km/h

11 A high-speed train has a top speed of 70 m/s. What is this speed in km/h?

12 **P / R** A swordfish can swim at up to 100 km/h.
  A cheetah can run at a top speed of 30 m/s. Which is faster? Explain your answer.

13 Use the formula $v = u + at$ to work out v when
  a $u = 5$, $a = 3$ and $t = 6$
  b $u = 15$, $a = 1.5$ and $t = 8$

14 Use the formula $v = u + at$ to work out u when
  a $v = 11$, $a = 2$ and $t = 4$
  b $v = 20$, $a = 2.5$ and $t = 6$

15 Use the formula $v = u + at$ to work out a when
  a $v = 9$, $u = 1$ and $t = 4$
  b $v = 12$, $u = 5$ and $t = 14$

16 Copy and complete the table using the formula $s = ut + \frac{1}{2}at^2$

s (m)	u (m/s)	a (m/s²)	t (s)
6		4	1
9	7		1
18		5	2
24	5		3

17 Copy and complete the table using the formula $v^2 = u^2 + 2as$

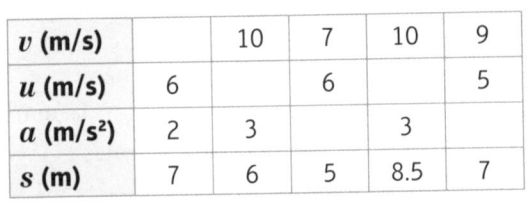

v (m/s)		10	7	10	9
u (m/s)	6		6		5
a (m/s²)	2	3		3	
s (m)	7	6	5	8.5	7

18 A car starts from rest and accelerates at $3\,\text{m/s}^2$ for 250 m.
   a Write down the values of u, a and s.
   b Using the formula in **Q17**, work out the final velocity in m/s.

14.5 Direct and inverse proportion

1 **R** The table shows the distance d travelled by a car over a period of time t.

Distance, d (km)	6	12	18	24	30
Time, t (minutes)	8	16	24	32	40

   a Write the ratios for the pairs of values in the table.
   b Is d in direct proportion to t? Explain your answer.
   c Write a formula that shows the relationship between distance d and time t.
   d Use your formula to work out the distance travelled after 20 minutes.

2 a Write each statement using the $\propto$ symbol.
      i P is proportional to I
      ii E is proportional to m
   b Now write each statement using k, the constant of proportionality.

Example

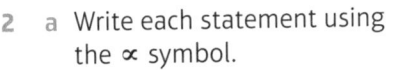

3 y is proportional to x.
   When $x = 5$, $y = 15$.
   a Write the statements of proportionality for y and x.
   b Use the given values of y and x to find k.
   c Find y when $x = 8$.

4 c is directly proportional to r.
   When $r = 8$, $c = 14$.
   a Write a formula for c in terms of r.
   b Find c when $r = 6$.

5 The number of pounds P is directly proportional to the number of dollars D.
   a One day $1 = £0.60$
      Write a formula connecting P and D.
   b Use your formula to convert $120 to £.

6 **R** X and Y are in inverse proportion. Copy and complete the table.

X	10	20	24		
Y	12			80	16

> **Q6 hint** XY = constant when X and Y are in inverse proportion.

7 Write down whether each of these equations represents direct proportion, inverse proportion or neither.
   a $y = 3x$
   b $x + y = 10$
   c $x - y = 2$
   d $y = 7 + x$
   e $xy = 8$

8 a Write each statement using the $\propto$ symbol.
      i s is inversely proportional to t.
      ii F is inversely proportional to d.
   b Now write each statement using k, the constant of proportionality.

9 y is inversely proportional to x.
   When $x = 8$, $y = 5$.
   a Write the statements of proportionality for y and x.
   b Use the given values of x and y to find k.
   c Find y when $x = 10$.

10 Exam-style question

   a is inversely proportional to b.
   $a = 50$ when $b = 0.8$
   Calculate the value of a when $b = 2.5$
   (3 marks)

11 The time t it takes to build a house is inversely proportional to the number of builders b.
   It takes 60 days for 8 builders to build a house.
   How many builders would be needed to build the house in 24 days?

14 Problem-solving

Solve problems using these strategies where appropriate:

- **Use pictures**
- **Use smaller numbers**
- **Use bar models**
- **Use x for the unknown**
- **Use a flow diagram**
- **Use more bar models**
- **Use formulae.**

1 The perimeter of this shape is 72 cm.
Work out the value of y.

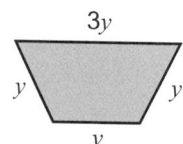

2 Micah puts a fence round the edge of his garden, which is a rectangle. He uses 24 m of fencing in total. The width of his garden is 4 m less than the length. How much fencing does Micah use for one of the longer sides of his garden?

> **Q2 hint** Create a formula using x for one side of the garden.

3 Georgie is shopping online. She is comparing prices between UK and USA stores.
The exchange rate is about 1.5 dollars ($) to the pound (£). The table shows the prices she found, including shipping fees.

Item	UK (£)	USA ($)
boots	125	145
make-up	35	62
jacket	147	200
bag	59	97

Georgie ordered each item from the country with the better deal.
Which country did she order each item from?

4 In 2011, VAT increased from 17.5% to 20%. Moussa bought a computer in 2015 for £380 + VAT.
a How much did the computer cost including VAT?
b How much cheaper would the computer have been with the previous VAT rate?

5 **R** Rita is selecting a sample of her employees. She selects the 10 employees who sit closest to her.
Is Rita taking a random sample of her employees? Explain your answer.

6 The total land area of Great Britain is 229 848 km². The total land area of the county of Yorkshire is 11 903 km².
a What percentage of the land area of Great Britain is covered by Yorkshire?
Give your answer to 3 significant figures.
b The county of Cumbria covers 2.94% of Great Britain's total land area. Approximately how many square kilometres is it?
Give your answer to 3 significant figures.

7 **R** Kayla and her friends are putting packets of pasta into packing boxes. A single packet of pasta has dimensions 15 cm × 20 cm × 6 cm. The packing boxes measure 45 cm × 40 cm × 60 cm.
a 12 packing boxes are filled to capacity with packets of pasta.
How many packets have they packed? Show how you worked it out.
b Jacob and his friends are using the same size packing boxes for packets of rice. A single packet of rice has dimensions 15 cm × 20 cm × 5 cm. 11 packing boxes are filled with packets of rice.
Which group of friends has packed more packets? Show how you worked it out.

8 **R** Raul is baking a cake. His recipe gives the ingredients in ounces but his scales only weigh in grams. He knows that 6 ounces is about 170 grams.
a What ratio would Raul need to calculate to convert the measurements from ounces to grams?
b Make a conversion chart that shows the conversions from ounces to grams for weights from 1 to 12 ounces.
c Explain why Raul's chart is not exact.

9 Amir thinks of a number. He adds $2\frac{1}{2}$ to his number. Then he adds $3\frac{4}{5}$ to that sum. Finally, he subtracts $1\frac{1}{4}$. Amir's final total is 9. What number did Amir start with?

10
> **Exam-style question**
>
> Michael wants to buy a flight to Spain. He finds a sale price of £143. The sale gives him 35% off the original price. What was the original price of the flight? **(3 marks)**

15 CONSTRUCTIONS, LOCI AND BEARINGS

15.1 3D solids

1 Here is a child's building brick.

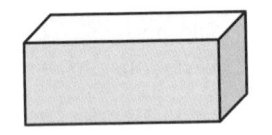

    a How many faces does the brick have?

    b How many edges does it have?

2 Exam-style question

Here is a cube.

Copy and complete each of the following sentences by inserting the correct number.

vertices

    a A cube has _____ faces.

    b A cube has _____ edges.

    c A cube has _____ vertices. **(3 marks)**

3 Here is a triangular prism.

    a What are the names of the shapes of its faces?

    b Write down the dimensions of each of the faces.

4 cm   4 cm   3 cm   6 cm

4 Here are some three-dimensional solids.

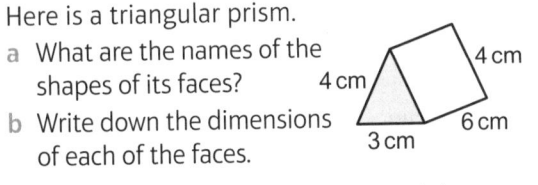

a   b   c   d   e

Match the correct name to each solid.

cylinder   sphere   cube   cone   cuboid

5 **R** Here are some three-dimensional shapes. They are all pyramids or prisms.

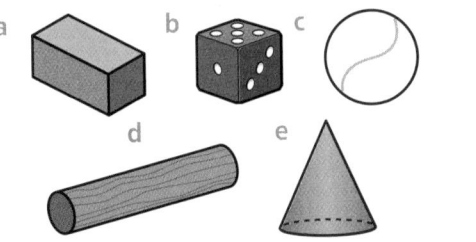

triangular prism   tetrahedron   octagonal prism

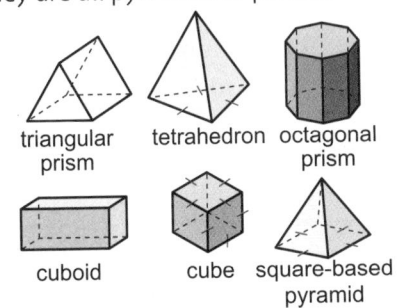

cuboid   cube   square-based pyramid

    a What shapes are the faces of the square-based pyramid?

    b What shapes are the faces of the octagonal prism?

    c Copy and complete the table.

Shape	Number of edges	Number of vertices	Number of faces
Cuboid	12	8	6
Cube			
Tetrahedron			
Octagonal prism			
Triangular prism			
Square-based pyramid			

    d Write a rule connecting the number of edges (E), vertices (V) and faces (F).

    e A 3D shape has 12 faces. Use your rule to work out the number of edges and vertices the shape has.

    f Does your rule work for all of the shapes in **Q4**?

> **Q5d hint** Add together the number of faces and vertices and compare this with the number of edges.
>
> **Q5e hint** A decagon has 10 sides, so the cross-section of a decagonal prism will have 10 sides.

6 **R** Jacob says, 'A 3D solid can only have fewer vertices than faces if it has curved edges.' Is he correct? Explain your answer.

15.2 Plans and elevations

1 Copy the diagrams and draw in all the planes of symmetry for each shape.

Example

a 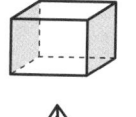   b

c   d

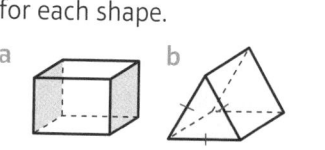

2 A rolled steel joist (RSJ) is moulded in this shape.

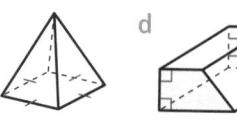

The RSJ is a prism and the cross-section of the RSJ is an H-shape.

The RSJ has three planes of symmetry. Copy the diagram and draw in the three planes of symmetry.

3 **R** A cake is made in the shape of a 5-pointed star.

How many ways can the cake be cut into equal halves?

4 Draw the plan view, the front elevation and side elevation of these cuboids on squared paper.

Example

a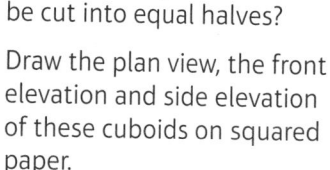
10 cm   1 cm   2 cm

b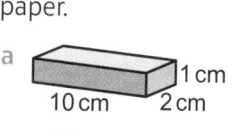
5 cm
2 cm   2 cm

c
3.5 cm
3 cm
9 cm

5 Here are the plan views of some solids. Name two possible solids that each could be.

a ▭   b ◯   c ▢

6 On squared paper, draw the plan, front elevation (shown by the arrow) and side elevation of this prism.

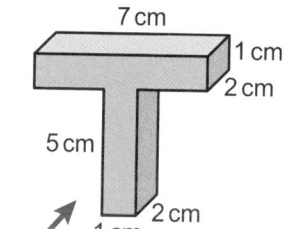
7 cm
1 cm
2 cm
5 cm
2 cm
1 cm

7 Here are the plan, front elevation and side elevation views of a prism, drawn on cm squared paper. Sketch the prism. Label its lengths.

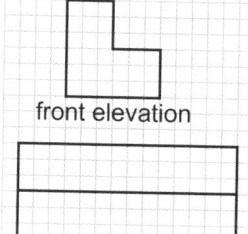
side elevation

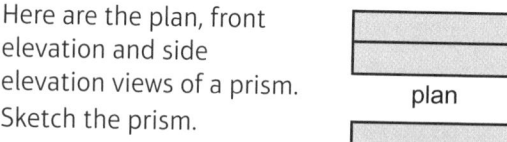
front elevation

plan

8 Here are the plan, front elevation and side elevation views of a prism. Sketch the prism.

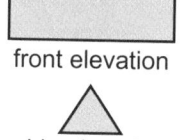
plan

front elevation

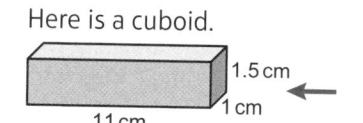
side elevation

9 Here is a cuboid.

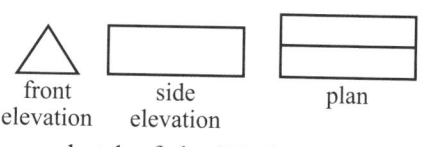
1.5 cm
1 cm
11 cm

Draw an accurate diagram of

a the side elevation (shown by the arrow)
b the plan view.

10 **Exam-style question**

Here are the front elevation, side elevation and plan of a 3D shape.

front elevation   side elevation   plan

Draw a sketch of the 3D shape. **(2 marks)**

11 This diagram shows a building in the shape of a prism.

Draw a sketch of the plan view of the building. Show the dimensions of the plan on your sketch.

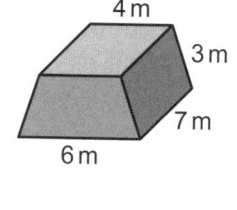
4 m
3 m
7 m
6 m

Q11 hint Imagine you are flying over the building directly above it. When you look down, what shape do you see? The shape of the top of the building or the shape of the bottom?

15.3 Accurate drawings 1

1 Make an accurate drawing of triangle ABC using a ruler and protractor.

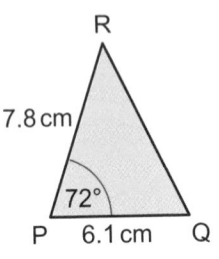

2 Make an accurate drawing of triangle PQR using a ruler and protractor.

Q2 hint Start by drawing and labelling the line PQ, and then mark the angle RPQ from point P using a protractor. Next, accurately draw the line PR from P through the marked point and label the end of this line R. Finally, join Q and R with a straight line to complete your triangle.

3

The diagram shows a sketch of isosceles triangle XYZ.

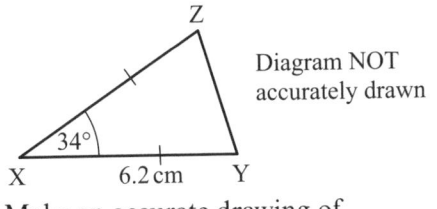

Diagram NOT accurately drawn

a Make an accurate drawing of triangle XYZ. **(2 marks)**

b Measure the length of side YZ. **(1 mark)**

4 **R a** Which of these five triangles are ASA triangles and which are SAS triangles?

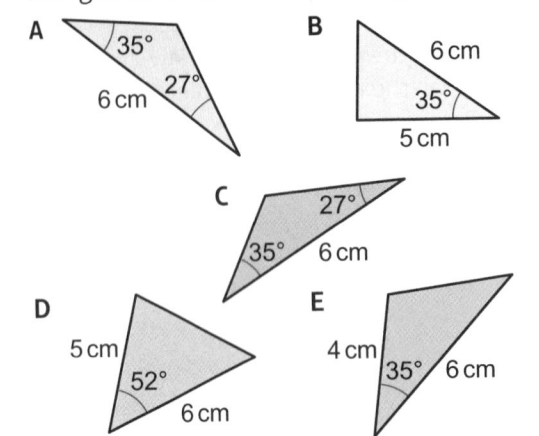

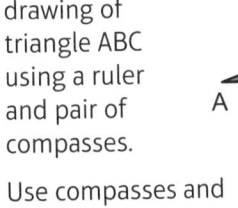

b Make accurate drawings of all the triangles using a ruler and protractor.

c Which two triangles are congruent?

5 Make an accurate drawing of triangle ABC using a ruler and pair of compasses.

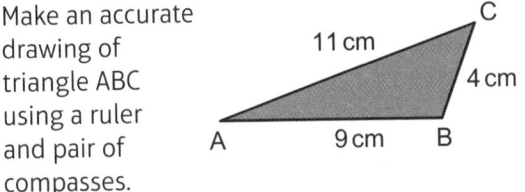

6 Use compasses and a ruler to draw an accurate diagram of this sketch.

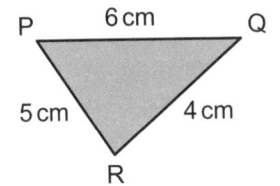

7 Make an accurate drawing of an isosceles triangle with two sides of length 4 cm and one side of length 6 cm.

8 **a** Make an accurate drawing of triangle LMN.

b How long is side LM?

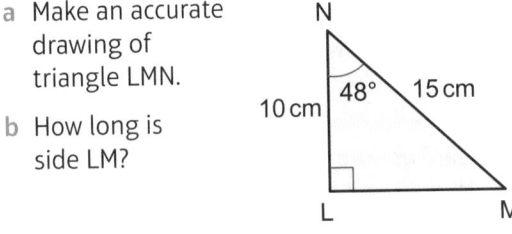

9 **R** Here is a sketch of an RHS triangle and an SSS triangle.

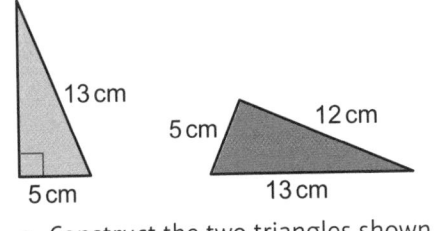

a Construct the two triangles shown.

b Are the triangles congruent?

15.4 Scale drawings and maps

1 On a scale drawing, 1 cm represents 5 miles.
   a What line length on the drawing represents
      i 10 miles
      ii 14 miles?
   b On the same drawing, what actual length do these represent?
      i 3 cm
      ii 4.2 cm

2 Here is a sketch of a plot of land.

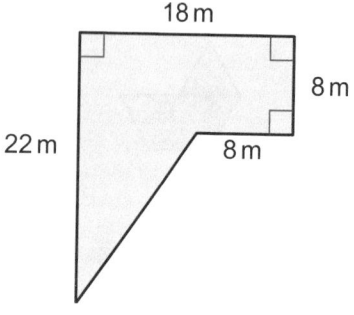

18 m

8 m

22 m

8 m

On cm squared paper, draw a scale diagram of the plot using the scale 1 cm represents 2 m.

3 A scale drawing of a rectangular garden has a scale of 1 : 50.

Example

   a The length of the garden in the plan is 40 cm. How long is the actual garden?
   b The width of the actual garden is 12 m. What is the width of the garden on the plan?

4 This diagram shows a scale plan of a swimming pool drawn on squared paper. The steps are 2 m by 3 m.

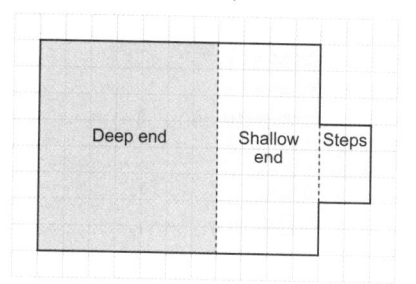

Deep end   Shallow end   Steps

   a Copy and complete the statement.
      1 cm on the diagram represents ☐ m.
   b Write the scale used as a ratio 1 : x.
   c Use the scale to work out
      i the dimensions of the shallow end
      ii the dimensions of the deep end
      iii the total surface area of the pool.
   d Copy the diagram and draw these items in the pool.
      i a large floating mat measuring 1 m by 2.5 m
      ii a slide measuring 1.5 m by 4 m

5 On a scale diagram, 4 cm represents a real-life length of 1.2 m.
   a Write the two lengths as a ratio of scale diagram : real life.
   b Write as a ratio in the form 1 : m.

> **Q5 hint** Both measurements should be in the same units.

6 Here is a map of part of Wales.
   The map has a scale of 5 cm : 100 km.

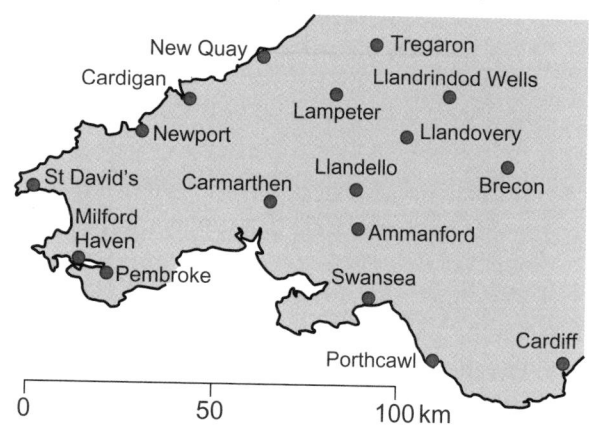

   a Using the map, work out the distance from
      i Cardigan to Swansea
      ii Swansea to Cardiff
      iii St David's to Brecon.
   b Which town is 19 km north of Swansea?

7 A landscape gardener makes a scale drawing of a garden using a scale of 1 : 20.
   She digs a flowerbed measuring 4 m by 0.5 m.
   a What are the dimensions of the flowerbed on the scale drawing?
   b A path on the scale drawing is 55 cm long. How long is it in real life?

8 **R** An estate agent draws a scale diagram of the upstairs of a house. The dimensions are

Bedroom 1   4 m by 3 m
Bedroom 2   3 m by 3 m
Landing   1.5 m by 4 m
Bedroom 3   3 m by 4 m
Bathroom   3 m by 4.5 m

All doors are 1 m wide.

a This scale diagram is incorrect. Explain why.

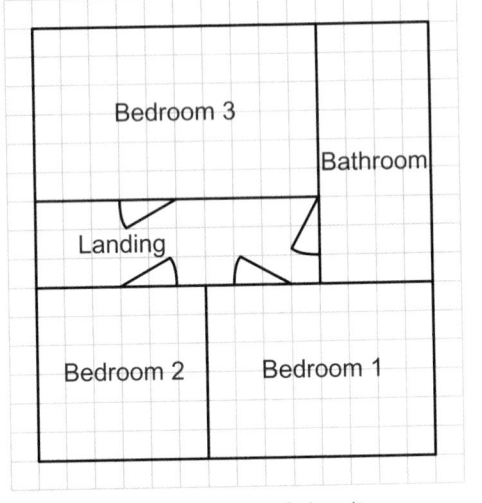

b Draw a correct version of the diagram on squared paper.

9 **R** An architect is designing an extension and wants to draw a scale diagram of it on a sheet of A3 paper.
The dimensions of the extension are 4 m by 7.5 m.
Which of these scales should the architect use to draw the scale diagram? Explain why.
A 1 : 20   **B** 1 : 200   **C** 1 : 10   **D** 1 : 5   **E** 1 : 100

10 The scale factor on a map is 1 : 250 000.
a What distance in metres does 10 cm represent?
b What distance in kilometres does 12 cm represent?
c How many centimetres on the map is a real distance of 35 km?

11 **R** The scale on an Ordnance Survey map is 1 : 50 000.
a What length on the map would a 12 km walk be?
b How many km would a walk represented by a length of 40 cm be?

12 **R** Three phone masts are arranged like this:
Mast A is 60 km south of Mast B.
Mast C is 45 km northwest of Mast B.
a Draw a scale drawing showing these three masts. Use a scale of 1 cm to 10 km.
Mast D is placed 65 km east of Mast A.
b Draw Mast D on your diagram.
c Use your diagram to find the distance from
   i   Mast C to Mast D
   ii   Mast A to Mast C.

15.5 Accurate drawings 2

1 **R** Which of these are not a net of a tetrahedron?

A 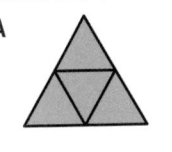 B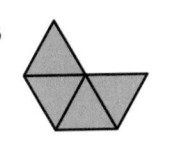

C   D
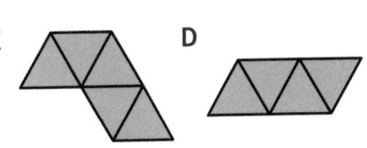

2 In triangle ABC, AB = 7.4 cm, BC = 5.9 cm and angle ABC = 62°.
a Draw a sketch of triangle ABC showing all three given measurements.
b Now make an accurate drawing of triangle ABC.
c Measure the size of angle CAB.

3 Using compasses, make an accurate drawing of triangle PQR where PQ = 9.2 cm, QR = 7.6 cm and PR = 3.1 cm.

4 Here is an isosceles trapezium, WXYZ.

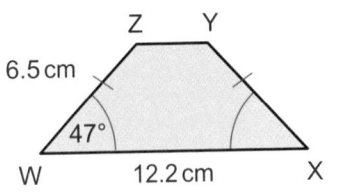

a Make an accurate drawing of this trapezium.
b Measure the length of side YZ.
c Measure the size of angle XYZ.

5 Follow these instructions to construct a regular octagon inside a circle.

a Draw a circle with radius 8 cm. Label the centre O.

b Draw a vertical line from O up to a point on the circumference. Label this point A.

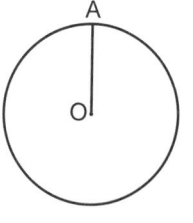

c To work out the angle you need to measure in the middle, divide 360° by the number of sides of an octagon. Call this angle x.

d Starting at line OA, measure angle x at the centre of the circle. Draw a line from O to the circumference through your measured angle. Label the new point on the circumference B and the angle x.

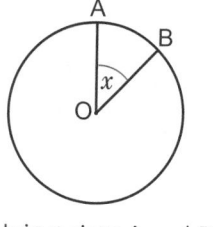

e Join points A and B together with a straight line. Line AB is the first side of your octagon.

f Repeat parts **d** and **e** until you have drawn the whole octagon.

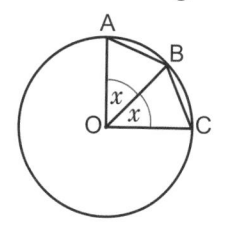

6 ABCD is a cyclic quadrilateral. O is the centre.
OA = OB = OC = OD = 7 cm.
Angle AOB = 112° and angle BOC = 76°.
Angle COD = angle DOA.

a Using a ruler, compasses and a protractor, draw polygon ABCD.

b Using your diagram, measure angle COD and the length of the line DA.

> **Q6 hint** A **cyclic quadrilateral** is a four-sided polygon whose vertices are all on the circumference of a circle.

7 On squared paper, draw an accurate net for this triangular prism.

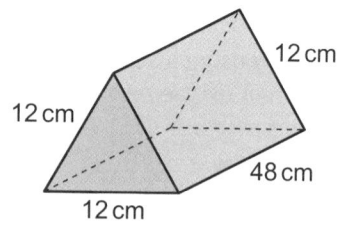

8 **P** A packaging company makes small boxes in the shape of a triangular prism.

Use a ruler and compasses to draw a scaled diagram of an accurate net for the box.
Use a scale of 1 cm to represent 3 cm.

15.6 Constructions

1 a Use a ruler and compasses to construct the perpendicular bisector of a line 13 cm long.

b Check, by measuring, that your line cuts the original line exactly in half.

Example

2 AB = 10 cm. C is 5 cm above the line AB.

a Construct the line AB and mark one possible position of C.

b Using a ruler and compasses, construct a perpendicular line to AB which passes through point C.

Example

3 a Construct triangle PQR where PQ = 5 cm, QR = 4 cm and RP = 2 cm.

b Mark a point X somewhere inside the triangle.

c Accurately construct the line that shows the shortest distance from X to the line PR. Measure this line.

4 **R** A, B and C are three railway stations.

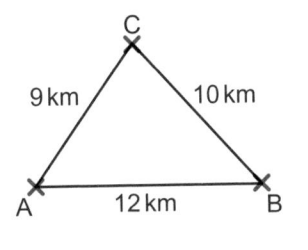

The railway company wishes to build a track from C to the line AB. The new track must be the shortest length possible.

a Draw the diagram using a scale of 1 cm to 2 km and construct on the new track using a ruler and compasses.

b How long will the new track be in km?

5 a Draw a horizontal line.

b Use your compasses to draw arcs on the line to find the midpoint of the line segment between the arcs.

c Construct the perpendicular bisector for this new line segment.

6 a Draw an angle of 80° using a protractor. Construct the angle bisector.

b Check, by measuring with a protractor, that your angle bisector cuts the angle exactly in half.

Example

7 Copy the diagrams. Then construct the bisector of angle B in each triangle.

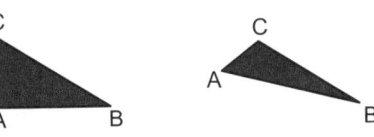

8 ⬢ **Exam-style question**

Copy this angle and then use a ruler and compasses to **construct** the bisector of the angle.

You must show all your construction lines. **(2 marks)**

Exam hint
Use the equipment they tell you to use. Using a protractor for this question gains no marks.

9 Draw a 125° angle and construct its angle bisector.

10 **R** How could you use your knowledge of constructing perpendicular bisectors and angle bisectors to find the exact height of this triangle?

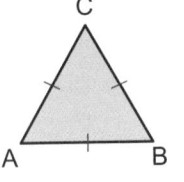

11 a Construct a right angle using a ruler and a pair of compasses.

b Bisect your right angle to give an angle of 45°.

15.7 Loci and regions

1 Mark a point A on a piece of paper. What is the locus of points that are 7 cm from point A?

2 **R** Two points Q and R are 6 cm apart.

a Draw points Q and R.

b A point S moves so that it is always equidistant (the same distance) from Q and R. Construct the locus of point S.

3 a Draw a line 10 cm long.

b A point moves so that it is always 3 cm away from the line segment. Construct the locus.

4 a Draw a 70° angle, PQR. Make each arm of the angle at least 5 cm long.

b Sketch the locus of points that are exactly the same distance from PQ and QR.

c Construct the locus accurately on your diagram using a ruler and compasses.

5 PQRS is an isosceles trapezium.

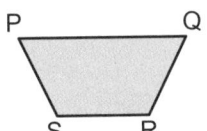

a Copy the diagram and draw the locus of points that are equidistant from P and Q.

b Draw the locus of points that are equidistant from the lines SR and SP.

c Mark the intersection of the loci with a cross X.

6 **R** Chewing gum is stuck on a bicycle wheel.

The wheel travels through 4 complete rotations before the gum comes off.

a Sketch the locus of points that the centre of the wheel moves as the wheel completes 4 rotations.

b Sketch the locus of points that the chewing gum moves before it falls off.

7 A rectangular piece of card is placed on a horizontal surface.

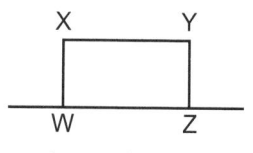

a The card is first rotated 90° clockwise about vertex Z. Copy or trace the diagram in colour 1. Draw the new position of the rectangle in one colour. Draw the locus of the vertex W in a second colour.

b The card is then rotated 90° clockwise about vertex Y and then 90° clockwise about vertex X. Draw the new positions of the card in your first colour and the remainder of the locus of vertex W in your second colour.

8 **R** Match each locus to the correct description.

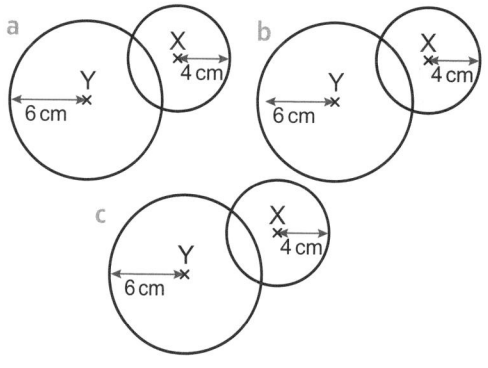

A The shaded region represents the area which is less than 4 cm from point X.

B The shaded region represents the area which is less than 6 cm from point Y.

C The shaded region represents the area which is less than 4 cm from X and less than 6 cm from Y.

9

> **Exam-style question**
>
> The diagram shows an accurate scale drawing of two radio transmitter masts.
>
>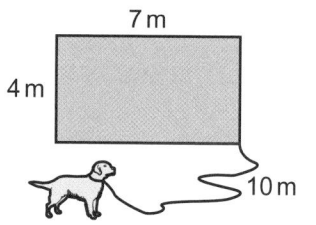
>
> Scale: 1 cm to 5 km
>
> Both masts can transmit up to 20 km.
>
> Copy the diagram and shade the region in which you can pick up a signal from both transmitters. **(3 marks)**

10 **P** A dog is tied by a 10 m rope to the corner of a shed which measures 7 m by 4 m.

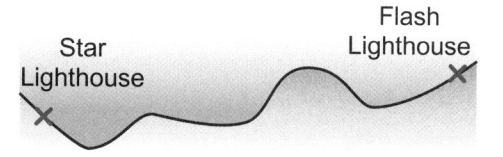

Copy the diagram and shade in the area where the dog can roam.

11 Two lighthouses are 120 km apart on a stretch of coastline.

A third lighthouse is being built exactly halfway between the two lighthouses on the coast.

Draw a scale diagram, using a scale of 1 cm : 12 km and mark the position of the new lighthouse.

> **Q11 hint** Construct the perpendicular bisector of the line joining the lighthouses.

12

The diagram shows a field.

Q   R

P   S

The scale of the diagram is 1 cm represents 20 m.

A sprinkler is being installed.

The sprinkler must be nearer to PS than QR and within 20 m of the point P.

Copy the diagram and shade the region where the sprinkler may be placed.

(3 marks)

15.8 Bearings

1 Write the bearing of B from A in these diagrams.

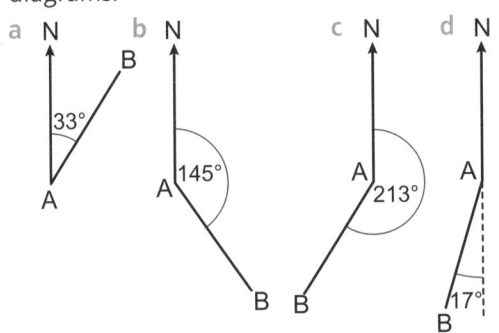

a N B 33° A

b N A 145° B

c N A 213° B

d N A 17° B

2 The bearing of B from A is 137°.
Work out the bearing of A from B.

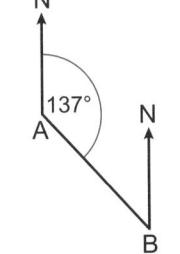

N 137° A N B

3

Work out the bearings of M from L.

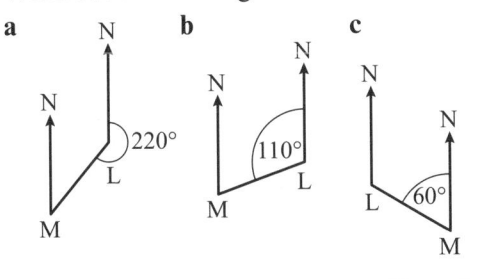

a N N 220° L M

b N N 110° L M

c N N L 60° N M

(3 marks)

4

Work out the bearing of

a Y from X

b X from Z.

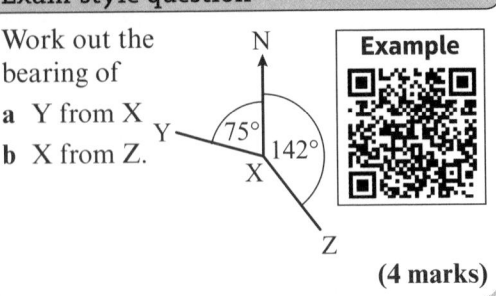

N 75° Y 142° X Z

Example

(4 marks)

5 Draw these bearings accurately. Use the scale 1 cm : 25 km.

a Petersfield is 200 km from Lottisham on a bearing of 035°.

b Steep is 150 km from Gunport on a bearing of 149°.

6 On a camp site Amy pitches her tent 300 m from the shower block on a bearing of 160°.
The kitchen is 200 m from the tent on a bearing of 75°.

a Draw an accurate diagram marking the positions of Amy's tent, the shower block and the kitchen. Use a scale of 1 cm : 50 m.

Use your diagram to find

b the distance between the shower block and the kitchen

c the bearing of the kitchen from the shower block

d the bearing of the shower block from the kitchen.

7 P Two lifeboat stations are at points A and B on a coastline. The stations are 20 km apart.
They both receive a distress call from a sinking ship.
The ship is on a bearing of 112° from A and 230° from B.

a Draw an accurate diagram of the lifeboat stations and mark the position of the ship. Use a scale of 1 cm : 2 km.

b Find the bearing of the lifeboat station at B from the ship.

c How far, to the nearest km, is the ship from the closer lifeboat station?

> **Q7a hint** Draw the lifeboat stations at A and B first, then draw in the bearings to the sinking ship. The ship's position will be where your two bearing lines intersect. Remember to use the scale carefully.

8 **P** Two boats A and B are racing to the finishing buoy. A is 500 m from B on a bearing of 320°. B is 300 m from the finishing buoy.

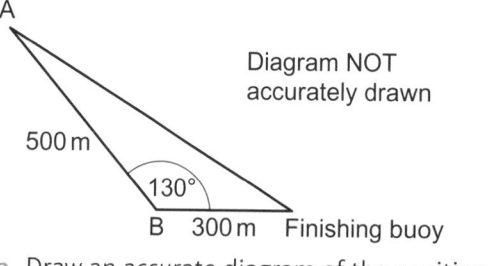

Diagram NOT accurately drawn

a Draw an accurate diagram of the positions of the two boats and the buoy.

b Use your diagram to find

  i the distance of boat A from the finishing buoy

  ii the bearing of the buoy from boat A.

9 The diagram shows the positions of a frigate (F) and a helicopter (H) relative to the port.

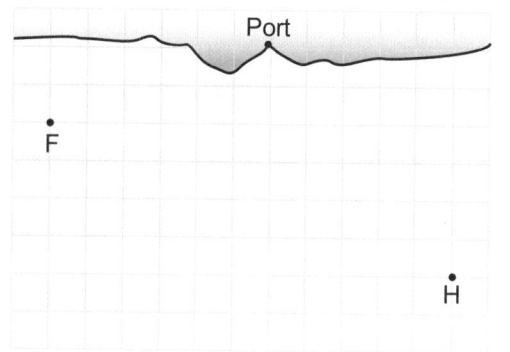

a What is the bearing of the frigate from the helicopter?

b What is the bearing of the helicopter from the port?

10 **P** This map shows the positions of two hospitals with Air Ambulance helicopters.

Scale: 1 cm represents 50 miles

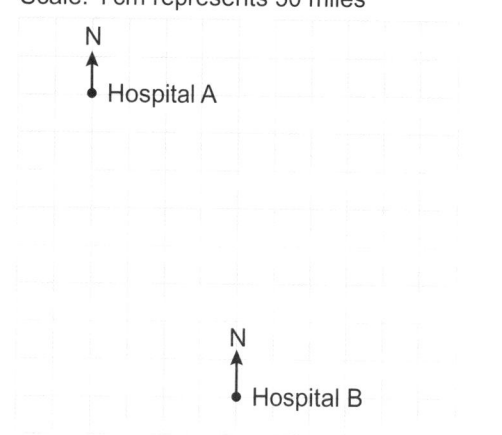

An injured climber is on a bearing of 112° from Hospital A and 37° from Hospital B.

The Air Ambulance helicopters respond to emergencies within a radius of 150 miles.

a Make an accurate scale drawing of the map.

b Mark the exact position of the climber on your diagram using the bearings.

c Show the region covered by each helicopter on your diagram.

d Which hospital should respond to the call?

15 Problem-solving

Solve problems using these strategies where appropriate:

Example

• **Use pictures**
• **Use smaller numbers**
• **Use bar models**
• **Use x for the unknown**
• **Use flow diagrams**
• **Use more bar models**
• **Use formulae**
• **Use arrow diagrams.**

1 Tony has to cut out some triangles for a stained glass window.

To fit the pattern, each triangle needs to have sides measuring 6 cm, 7 cm and 8 cm.

What are the angles in the triangles that Tony needs to make?

2 Ann is given the front elevation, side elevation and plan of a wooden box she has been asked to make.

She needs to see what shape the box will be when it is all put together.

10 cm
front elevation

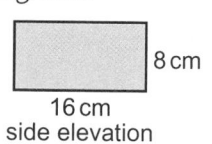
8 cm

16 cm
side elevation

16 cm

10 cm

plan

Use the drawings to draw the finished box on cm squared paper.

3 **R** The ruins of a stone circle were found. The circle was 98 yards in diameter. Each stone was 1 yard wide and there was a gap of 13 yards between each pair of stones.

a Explain how a historian could make a scale drawing of the stone circle and work out where any missing stones would have stood.

b How many stones would have stood in the complete ring?
Explain how you worked out your answer.

4 Jess starts at point X and takes his fishing boat out to his usual fishing place at point P. He doesn't catch many fish, so he decides to try an area a friend told him about at point Q.

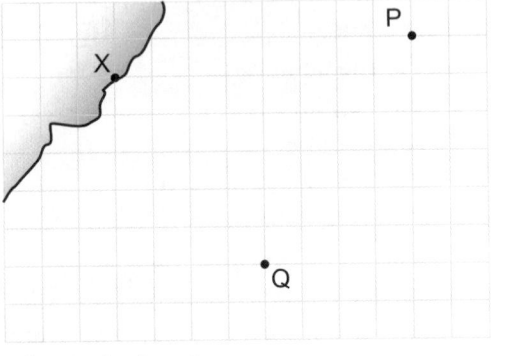

What is the bearing

a of P from X

b of Q from P?

5 **R** Siyona has a bag of four coins – a 2p, 10p, 20p and 50p. She takes one coin from the bag at random and flips it.

a What are all the possible outcomes of this event?

b What is the probability of Siyona taking a 20p coin and it landing heads?

c What is the probability of her taking a coin greater than 10p and it landing tails?

d Describe two events involving picking these coins that would have a probability of $\frac{1}{6}$.

6 **R** Craig wants to decorate his bedroom before his parents return home from holiday. He has 2 hours to get it painted.
It takes 3 people 5 hours to paint the room.

a Explain how Craig can work out how many people he would need to get it done in 2 hours.

b How many people does he need? Why does Craig need to round up his answer?

Q6 hint Could an arrow diagram help?

7

Exam-style question

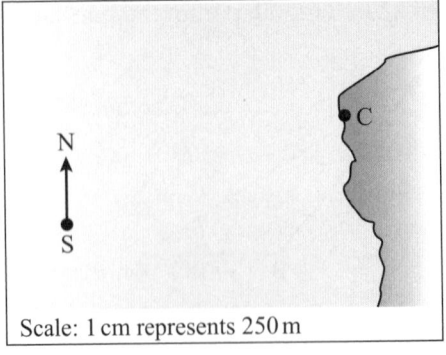

Scale: 1 cm represents 250 m

The position of a ship, S, is marked on the map. Point C is on the coast.

Ships must not sail closer than 500 m to point C.

The ship sails on a bearing of 037°

Will the ship sail closer than 500 m to point C?

Use a construction to justify your answer.
(3 marks)

June 2013, Q13, 1MA0/1H

8 **R** Oscar books a function room for a party. The cost to hire the room is £40 per hour (h) plus £12 per head for each guest (g).

a Write an expression for the cost of hiring the room.

Oscar has a budget of £500. The room is booked from 11.30 am until 3.30 pm.

b What is the maximum number of guests Oscar can invite without exceeding his budget?

9 Logan is on a scavenger hunt. He has found the first item at point A and has been given information about the next item at point B. Point B is on a bearing of 125° from point A at a distance of 1 km.

a Draw a diagram to show points A and B.

At point B Logan is given information about point C. It is on a bearing of 171° from point B at a distance of 1.5 km.

b Mark point C on your diagram.

c How far is point C from point A?

10 Carly is planning where she can put a pond in her garden. She wants it no further than 16 feet from the house and it needs to be in the shade of a tree, which offers shade 4 feet around it. The tree is 18 feet from the house. Draw a sketch map to show a possible area for the pond.

16 QUADRATIC EQUATIONS AND GRAPHS

16.1 Expanding double brackets

1 Write an expression for the area of each rectangle.

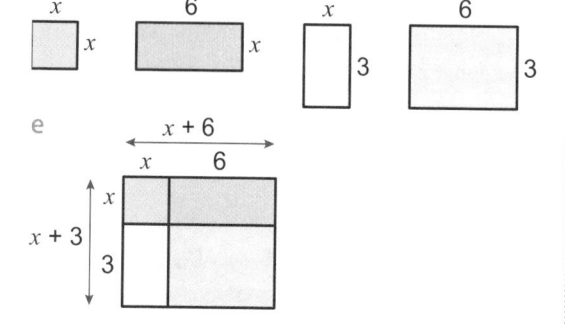

a   b   c   d

e

2 **R** i Write an expression for the area of each small rectangle.
   ii Write an expression for the area of the large rectangle.
   Collect like terms and simplify.

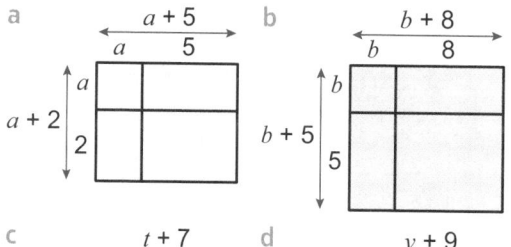

a   b

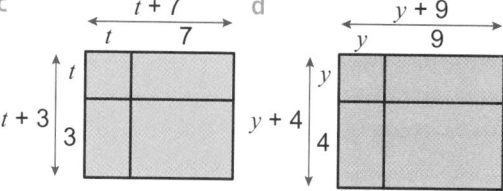

c   d

3 Expand and simplify
   a $(x + 1)(x + 3)$
   b $(h + 2)(h + 4)$
   c $(d + 5)(d + 6)$
   d $(p + 7)(p + 4)$
   e $(z + 10)(z + 6)$
   f $(y + 4)(y + 11)$

Example

4 Expand and simplify
   a $(x + 3)(x - 1)$
   b $(n - 2)(n + 3)$
   c $(k - 6)(k + 2)$
   d $(a + 3)(a - 7)$
   e $(y - 1)(y - 5)$
   f $(z - 2)(z - 6)$

> **Q4 hint**
>
> $3 \times -1 = \square$ $1 \times x = \square$ $-2 \times -6 = \square$

5 What is the missing term?
   $(x + 6)(x + \square) = x^2 + 10x + 24$

6 **R** Ryan says that the expansion of $(x + 2)(x - 5)$ is the same as $(x - 5)(x + 2)$.
   Is he correct? Give reasons for your answer.

7 **R** Tyler expands and simplifies $(x - 2)(x - 8)$.
   He says that the answer is $x^2 + 10x - 16$.
   Is Tyler correct? Give reasons for your answer.

8

> **Exam-style question**
>
> a Expand $6(x + 3)$ **(1 mark)**
> b Expand $2y(5y + 3)$ **(2 marks)**
> c Expand and simplify $(p + 7)(p + 2)$
>   **(2 marks)**

9 **R** Which of these are quadratic expressions?
   Give reasons for your answers.
   a $x^2 + 4x - 7$
   b $p^2 + 3p$
   c $7y + 12$
   d $z^3 - z^2 + 10$
   e q^2
   f $20 - 5y - y^2$

10 Expand and simplify
   a $(x + 4)^2$ b $(b + 3)^2$
   c $(d - 2)^2$ d $(k - 5)^2$

11 Square these expressions.
   Simplify your answers.
   a $y + 5$ b $z + 10$
   c $m - 1$ d $n - 3$

12 Match the cards that have equivalent expressions.

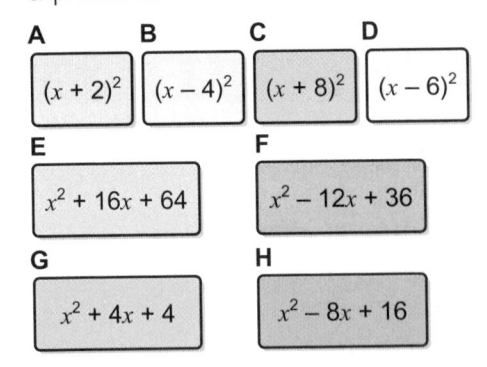

A $(x + 2)^2$	**B** $(x - 4)^2$
C $(x + 8)^2$	**D** $(x - 6)^2$

E $x^2 + 16x + 64$

F $x^2 - 12x + 36$

G $x^2 + 4x + 4$

H $x^2 - 8x + 16$

13 Exam-style question

The length of a side of a square tile is $a + 10$.

$a + 10$

[square diagram with sides labelled $a + 10$]

$a + 10$

a Work out an expression for the area of the tile.
Simplify your answer. **(2 marks)**

b Five tiles are used in each row on a wall.
Write an expression for the total area of a row of tiles.
Simplify your answer. **(1 mark)**

14 P The length of a rectangle is $x + 7$ and its width is $x + 5$.
Show that the area of the rectangle is $x^2 + 12x + 35$.

15 R Jasmine expands and simplifies $(x + 3)(x - 3)$.
She says that the answer is $x^2 - 9$ and this is a quadratic expression.
Is she correct? What do you notice about the two terms in the answer?

16.2 Plotting quadratic graphs

1 a Copy and complete the table of values for $y = x^2$ for values of x from -4 to 4.

x	-4	-3	-2	-1	0	1	2	3	4
y									

b Plot the graph of $y = x^2$.
Join the points with a smooth curve.
Label your graph.

Q1a hint Work out each y-value by substituting each x-value into the equation, e.g. when $x = -3$, then $y = x^2 = (\Box)^2 = \Box$

2 a Copy and complete the table of values for $y = x^2 + 2$.

x	-3	-2	-1	0	1	2	3
x^2	9			0			
$+2$	$+2$			$+2$			
y	11			2			

b Plot the graph of $y = x^2 + 2$.
Join the points with a smooth curve. Label your graph.

3 R a Copy and complete the table of values for $y = x^2 - 2x + 3$.

x	-2	-1	0	1	2	3	4
x^2	4						
$-2x$	$+4$					-4	
$+3$	$+3$						$+3$
y	11						

b Plot the function $y = x^2 - 2x + 3$.

c What do you notice about the y-values in the table?

4 For each graph, write down
  i the equation of the line of symmetry
  ii the turning point
  iii the y-intercept.

Example

a
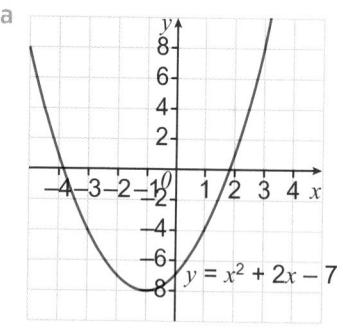
$y = x^2 + 2x - 7$

b
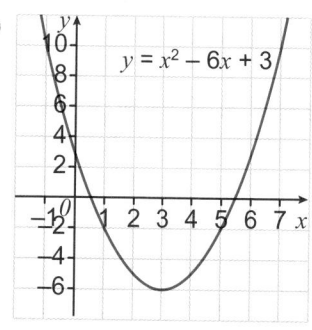
$y = x^2 - 6x + 3$

5 Copy and complete the table and plot the function $y = -2x^2$.

x	−3	−2	−1	0	1	2	3
x^2	9				1		
y	−18				−2		

6 **R** Which of these are graphs of quadratic functions?

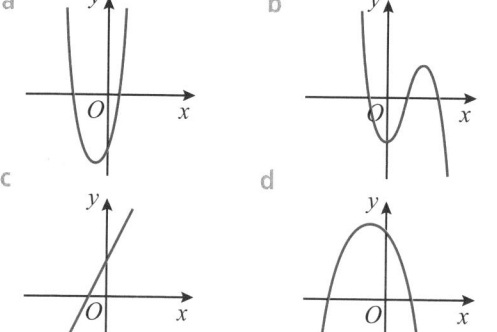

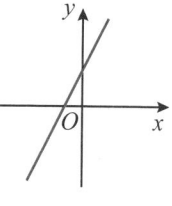

7 **R / P** The graph shows the height of a cricket ball against time.

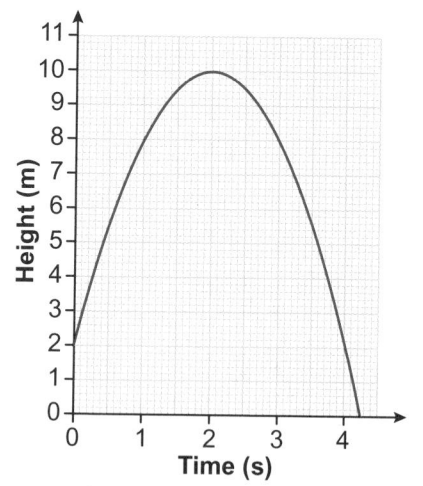

a i What is the maximum height that the cricket ball reaches?
  ii At how many seconds does it reach its maximum height?
b At what times does the ball reach 7 m?
c When does the ball hit the ground?

8 **R** The graph $y = x^2$ can be used to work out the area of a type of metal sheet, where the y-axis represents area in m², and the x-axis represents side length in m.
a Use your graph of $y = x^2$ from **Q1** to work out the area of a metal sheet with side length
  i 2 m
  ii 3 m
  iii 1.5 m.
b What shape is the metal sheet? Why?

9 **P** A firework is launched from the ground. The table shows the height h of the firework and the time t that it takes to travel.

Time, t (s)	0	1	2	3	4
Height, h (m)	0	26	48	66	80

a Plot a graph of the height of the firework against time.
b From your graph, estimate how long the firework takes to reach 40 metres.
c From your graph, estimate the height of the firework at 0.5 seconds.

16.3 Using quadratic graphs

1 Look at the graph of $y = x^2 - 4x - 5$.

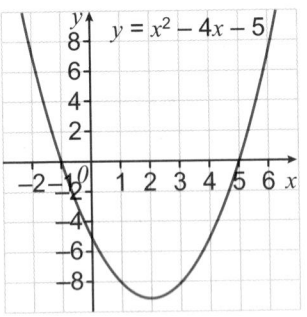

$y = x^2 - 4x - 5$

a Write down the coordinates of the points where the graph crosses the x-axis.

b Write down the roots of $x^2 - 4x - 5 = 0$.

2 Use the graphs to solve the equations.

a $x^2 + 4x - 12 = 0$

b $x^2 - 2x + 1 = 0$

c $x^2 - 3x - 4 = 0$

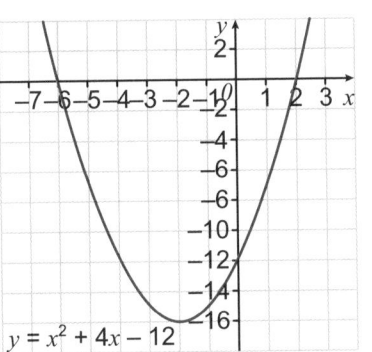

$y = x^2 + 4x - 12$

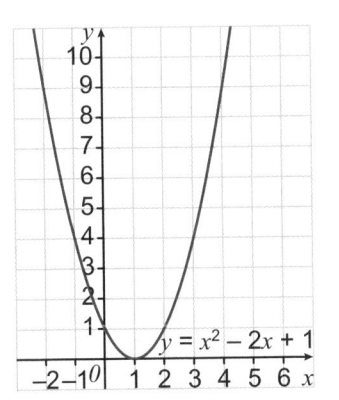

$y = x^2 - 2x + 1$

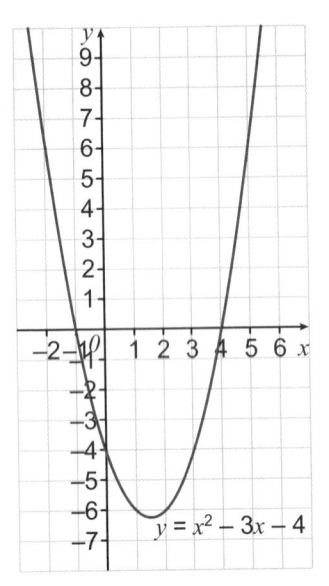

$y = x^2 - 3x - 4$

3 a On graph paper, draw the graph of $y = x^2 + 2x - 5$ from $x = -4$ to $x = 2$.

b From your graph, estimate the solutions to $x^2 + 2x - 5 = 0$.

4
Exam-style question

a Copy and complete the table of values for $y = x^2 + 3x - 2$ **(2 marks)**

x	-4	-3	-2	-1	0	1	2
y	2	-2				2	

b Draw the graph of $y = x^2 + 3x - 2$ for values of x from $x = -4$ to $x = 2$ **(2 marks)**

c Find estimates of the solutions of the equation $x^2 + 3x - 2 = 0$ **(2 marks)**

5 Use these graphs to solve the equation $x^2 + 4x = 5$.

Example

$y = 5$

$y = x^2 + 4x$

6 Use these graphs to solve the equation $x^2 + x - 9 = -3$.

$y = x^2 + x - 9$

$y = -3$

7 **a** On graph paper draw the graph of $y = x^2 + x - 4$ for values of x between -3 and 3.

b Use the graph to estimate the solutions to $x^2 + x - 4 = 0$.

c Use the graph to estimate the solutions to $x^2 + x - 4 = -1$.

8 Use the graph to solve the equations.

a $-x^2 + 4x = 0$

b $-x^2 + 4x = 2$

$y = -x^2 + 4x$

9

a Copy and complete the table of values for $y = x^2 + 3x$

x	-4	-3	-2	-1	0	1	2
y			-2		0	4	

(2 marks)

b Draw the graph of $y = x^2 + 3x$ for values of x from $x = -4$ to 2. **(2 marks)**

c Solve $x^2 + 3x + 4 = 2$ **(2 marks)**

Q9c hint Rearrange $x^2 + 3x + 4 = 2$ so that one side of the equation is the same as $x^2 + 3x$, then use your graph.

10 R a Copy and complete the table of values and plot the function $y = x^2 + 3x - 2$.

x	-5	-4	-3	-2	-1	0	1	2
y								

b Use your graph to solve the equation $x^2 + 3x - 2 = 2$.

c Why are there no solutions to the equation $x^2 + 3x - 2 = -5$?

11 Write down the y-intercept of the function in **Q7**.

16.4 Factorising quadratic expressions

1 Copy and complete

a $(\square + 3)(x + 5) = x^2 + 8x + 15$

b $(x + \square)(x + 7) = x^2 + 9x + 14$

c $(x + 4)(x - \square) = x^2 + 2x - 8$

d $(x - \square)(\square - 2) = x^2 - 7x + 10$

2 Factorise

a $x^2 + 7x + 6$

b $x^2 + 9x + 20$

c $x^2 + 9x + 18$

d $x^2 + 10x + 16$

Example

3 Factorise

a $x^2 + x - 12$

b $x^2 + 3x - 4$

c $x^2 - 2x - 15$

d $x^2 - 4x - 12$

e $x^2 - 7x + 10$

f $x^2 - 9x + 14$

4 **Exam-style question**

Factorise $x^2 - 3x - 28$ **(2 marks)**

5 **P** An expression for the area of a square tile in cm² is $x^2 + 12x + 36$.
  a What is the length of a side of the tile?
  b Work out the area when $x = 5$.

6 Factorise
  a $x^2 + 10x + 24$ b $x^2 + 9x - 22$
  c $x^2 - 3x - 10$

7 **P** An expression for the area of a rectangle is $x^2 - 6x + 8$.
  Work out expressions for the length and height of the rectangle.

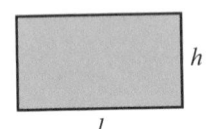
h
l

8 Expand and simplify
  a $(x + 1)(x - 1)$ b $(x + 5)(x - 5)$
  c $(x - 3)(x + 3)$ d $(x + 8)(x - 8)$

9 **R** Write two different expressions that are the difference of two squares using the terms in the box.

| x^2 | -25 | $+8$ | $+16$ | -9 | -20 | y | y^2 |

10 **P** Lucy cuts a small square of paper from a larger square.
  a Write an expression for the remaining area of the large square.
  b What type of expression is your answer to part **a**?

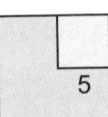
5
y

11 **Exam-style question**

  a Expand and simplify $(x + 4)(x - 5)$
                                          (2 marks)
  b Factorise $y^2 - 64$ **(1 mark)**

16.5 Solving quadratic equations algebraically

1 Copy and complete to solve $x^2 - 16 = 0$
  $x^2 - 16 + \square = 0 + \square$
  $x^2 = \square$
  $x = \square$ or $x = -\square$

2 Solve
  a $x^2 - 4 = 0$ b $y^2 - 49 = 0$

3 Solve
  a $x^2 - 5 = 20$ b $x^2 - 11 = 53$
  c $x^2 + 32 = 132$

Q3 hint Rearrange so you only have x^2 on the left-hand side.

4 Solve by factorising
  a $x^2 + 12x + 20 = 0$
  b $x^2 - x - 20 = 0$
  c $x^2 - 13x + 12 = 0$

Example

5 Solve
  a $x^2 + 4x + 4 = 0$ b $x^2 + 8x + 15 = 0$
  c $x^2 + 6x - 16 = 0$ d $x^2 - 2x - 8 = 0$
  e $x^2 - 7x + 12 = 0$ f $x^2 - 8x + 7 = 0$

6 **R** Match each equation to the correct solution.

A
$x = 6$

B
$x = 8$ and $x = -8$

C
$x = 4$ and $x = 5$

D
$x = -5$ and $x = -6$

E
$x^2 + 11x + 30 = 0$

F
$x^2 - 64 = 0$

G
$x^2 - 12x + 36 = 0$

H
$x^2 - 9x + 20 = 0$

7 **Exam-style question**

  a i Factorise $x^2 - 5x - 24$
    ii Solve the equation $x^2 - 5x - 24 = 0$
                                          (3 marks)
  b Factorise $x^2 - 81$ **(1 mark)**

8 Solve by factorising.
  a $x^2 - 25 = 0$
  c $t^2 - 1 = 0$
  b $a^2 - 100 = 0$
  d $r^2 - 144 = 0$

9 **P** Write a quadratic equation with solution $x = 2$ and $x = 6$.

10 a Plot the graph of $y = x^2 + 4x + 3$ for values of x from -4 to $+2$.
   b Use your graph to solve the equation $x^2 + 4x + 3 = 0$.
   c Solve $x^2 + 4x + 3 = 0$ by factorising. Show that the solutions match the solutions from your graph.

16 Problem-solving

Solve problems using these strategies where appropriate:

Example

- **Use pictures**
- **Use smaller numbers**
- **Use bar models**
- **Use x for the unknown**
- **Use flow diagrams**
- **Use more bar models**
- **Use formulae**
- **Use arrow diagrams**
- **Use graphs.**

1 A bag contains 40 stamps and 8 envelopes. What is the ratio of stamps to envelopes in its simplest form?

2 **R** Paul says that $y = 2x + 5$ is a quadratic equation.
   Explain why you think Paul is correct or incorrect.

3 **R** Ten friends play a computer game. Each friend plays each other friend once.
   a How many games are played in total?
   b Write an expression for the number of games played by m friends.

4 Jodie has a choice of getting the bus, the train or a taxi to travel to the airport.

Transport	Average speed	Distance
Bus	45 mph	24 miles
Train	65 mph	27 miles
Taxi	50 mph	21 miles

   Work out the time for each mode of transport to 1 decimal place.
   Which mode of transport will get Jodie to the airport the quickest?

5 **R** Rosa is buying a new TV. A TV is measured in inches from corner to corner. She has a space that can fit a TV no taller than 20 inches and no wider than 36 inches. The choices of TV size are 26", 32", 34", 40", 42" and 46".
   What is the largest sized TV Rosa can fit in her space?

6
   a Expand $3(x - 6)$ **(1 mark)**
   b Factorise $5y - 10$ **(1 mark)**
   c Expand and simplify
      $3(4w + 1) - 5(3w - 2)$ **(3 marks)**

7 Lucas drops a parachute toy from the second floor window of his house.
   He records the time at each metre it falls.
   His put his results in a table.

Height, h (m)	Time, t (s)
7	0
6	0.75
5	1.5
4	2.25
3	3
2	3.75
1	4.5
0	5.25

   a Plot a graph to show the rate of fall.
   b From your graph estimate how long the parachute would take to fall to 4.5 m above the ground.

8 Xavier knows that the area of a rectangular courtyard is $x^2 + 5x + 6$.
   Work out the expressions for the length and width of the courtyard.

   Q8 hint Can you use a picture or diagram to help you?

9 **R** Cho has been asked to put the equation $y = x^2 + 3x + 2$ into a graph.
   a What can she use to find the y-values for the graph?
   b Find the y-values for x-values from −3 to 3.
   c Plot the graph of $y = x^2 + 3x + 2$. Join the points with a smooth curve.

10 **R** Melissa is decorating a square picture frame. The width of the glass area in the middle is 6 inches. The total width of the frame is z inches.
   Work out an expression for the area of the frame.

17 PERIMETER, AREA AND VOLUME 2

17.1 Circumference of a circle 1

1 a Describe in words the relationship between the diameter of a circle and the radius.
   b The diameter of a circle is 12 cm. What is the radius?
   c The radius of a wheel is 13 cm. What is the diameter?

2 **R** The table shows the diameter and circumference of some everyday objects.

Object	Diameter, d (cm)	Circumference, C (cm)
plate	20	62.8
trampoline	250	785.4
drum	55	172.8
table	120	377.0

   a Work out the value of $\frac{C}{d}$ for each object. Round your answers to 2 d.p.
   b Copy and complete the formula for working out the circumference of a circle.
      $C = \pi \times \square$
   c A round table has diameter 1.5 m. Copy and complete the calculation to find the circumference.
      $C = \pi \times \square = \square$
      Round your answer to 1 decimal place.

3 Work out the circumference of each circle. Use the π button on your calculator. Round your answers to 1 d.p.

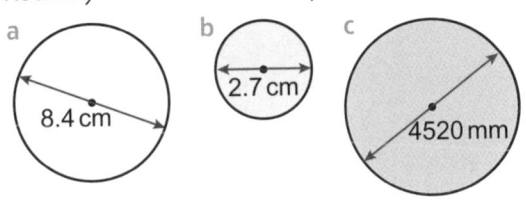

a 8.4 cm   b 2.7 cm   c 4520 mm

4 Work out the circumference of each circle. Use the π button on your calculator. Round your answers to 1 d.p.

a 2.5 cm   b 9.21 m   c 3.15 mm

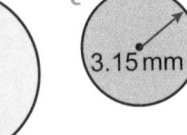

5
   A Ferris wheel has radius 28 m.
   Work out the circumference of the wheel giving your answer to 3 significant figures. **(2 marks)**

6 **P** Cheng is making labels for jam jars.
   Each jam jar has diameter 6 cm.
   She wraps the label all round the jar and allows an extra 3.5 cm of paper for gluing.
   a What length of paper does she need for one jam jar?
   b How many jam jars will she be able to label with a 1 metre strip?

7 **P** A tractor wheel has a radius of 3 feet.
   How far does the tractor travel for one revolution of the wheel?

8 The London Eye has a radius of 67.5 m.
   It rotates 34 times each day.
   What distance does one of the pods on the London Eye travel in a day?
   Give your answer in km to 3 s.f.

   Q8 hint Start by finding the distance travelled when the Eye makes one revolution.

17.2 Circumference of a circle 2

1 a What is the smallest value that can be rounded up to
      i   7 m to the nearest m
      ii 15 cm to the nearest cm
      iii 12 mm to the nearest mm?
   b Write a value that rounds down to
      i   12 cm to the nearest cm
      ii 35 km to the nearest km
      iii 107 mm to the nearest mm.

2 Martha is 127 cm tall to the nearest cm. Write down the smallest height she could be.

3 Rudi weighs 72.642 kg to the nearest gram.

   a What is his minimum possible weight in kilograms?

   b What is his maximum possible weight in kilograms?

   c Write an inequality to show the possible weights.

Q3 hint

$\square \leqslant w < \square$

4 The population of the UK is 64 000 000 to the nearest million.

Write an inequality to show the possible minimum and maximum populations.

5 Write an inequality to show the possible values for

   a 7.21 (rounded to 2 d.p.)

   b 200 (rounded to the nearest 100)

   c 19.7 (rounded to 3 s.f.)

   d 3200 (rounded to 2 s.f.)

6 **Exam-style question**

An airline charges extra for any suitcases weighing more than 22 kg.

Sophie's case weighs 22 kg to the nearest kg. Explain why she may still be charged. **(2 marks)**

7 **a** π = 3.141 592 654. Round π to 1 s.f.

   b Use the value of π from part **a** to work out the circumference of a circle with diameter 11 cm.

8 Work out the circumference of each of these circular objects. Round your answers to an appropriate degree of accuracy.

   a A cake with diameter 22 cm.

   b A lipstick with diameter 0.9 cm.

   c A sombrero with radius 0.45 m.

9 A circular helicopter landing pad has a radius of 15.4 m.

   a Write the circumference in terms of π.

   b Estimate the circumference of the landing pad.

   c Calculate the circumference to the nearest cm.

10 A circle has circumference 18π cm. Write down

   a the diameter   **b** the radius.

11 Work out, to 1 d.p., the diameter of a circle with circumference

   a 12.4 m

   b 18 cm.

12 Work out, to 3 s.f, the radius of a circle with circumference

   a 25 cm

   b 5.9 km.

13 **R** The circumference of a circle is 15 cm correct to the nearest cm.

   a Write the range of possible values for the circumference.

   b What are the possible values for the radius, correct to 1 d.p.?

17.3 Area of a circle

1 Work out the area of each circle. Round your answers to an appropriate degree of accuracy.

Example

   a

3 cm

   b
0.5 cm

   c
12.6 cm

   d
4.2 mm

2 **a** Estimate the area of each of these circular objects.

     i The head of a pin with a radius of 0.5 cm.

     ii A circular pond with a diameter of 8 feet.

     iii A crop circle with diameter 600 m.

   b Calculate each area in part **a** leaving your answers in terms of π.

   c Calculate each area in part **a** correct to 2 s.f.

3 **R** A circle has radius 3 cm. What is the correct area?

   A 6π cm^2 **B** 3π cm^2 **C** 9π cm^2 **D** 36π cm^2

   Explain your reasoning.

4 A circular flower bed has a radius of 2.4 m.
  a Work out the area of the flower bed.
  b A gardener plants 25 bulbs per square metre. How many bulbs does he plant?

5 A blob of ink is dropped onto a piece of blotting paper.
The ink expands to a circular shape with diameter 3.2 cm.
  a Work out the area of the circle.
  b The blotting paper measures 7 cm by 8 cm. What area of paper is not covered in ink?

6 For each circle work out
  i the radius
  ii the diameter.

Example

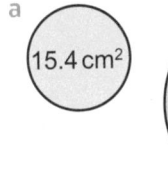 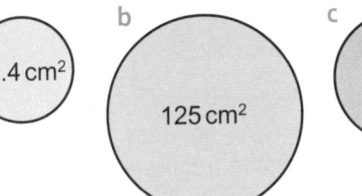

a 15.4 cm²   b 125 cm²   c 0.9 m²

7 A circular mosaic covers an area of 32.4 m².
What is the radius of the mosaic?
Give your answer to an appropriate degree of accuracy.

> **Q7 hint** Sketch a diagram.

8 A dog is attached to a post with a rope.

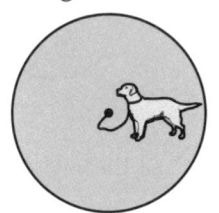

The dog has an area of 19.6 m² in which to play.
How long is the rope?

9 A door has a circular hole cut into it for a window.
The door is 2 m high and 0.95 m wide.
The window has diameter 0.5 m.
What is the area of door left to be painted?

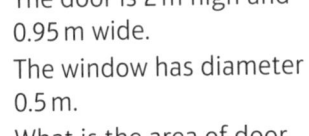

10 P Here is a wooden picture frame.
Work out the area of the frame correct to 3 s.f.

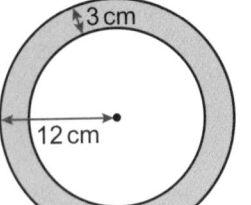

11

An ornamental garden is rectangular with three circular ponds within it.

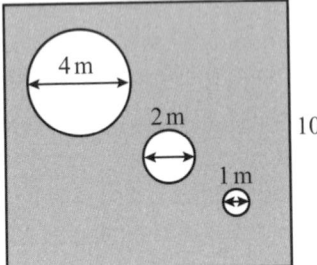

The rest of the garden is to be covered in turf.
Turf costs £12.99 per square metre.
Work out the cost of the turf.
Show all your workings.   **(5 marks)**

17.4 Semicircles and sectors

1 On this diagram what colour is the
  a chord
  b arc
  c sector
  d segment
  e tangent?

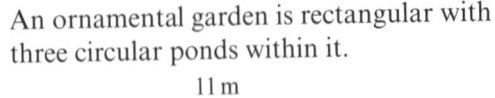

2 Work out the area of each semicircle.
Give your answers
  i in terms of π
  ii correct to 3 s.f.

a

b

9.4 cm   21.3 cm

3 Work out the area of each quarter circle. Give your answers
  i in terms of π
  ii correct to 1 d.p.

a 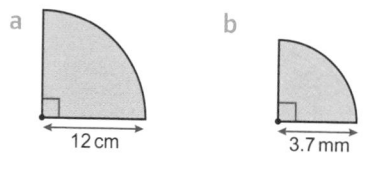 b
12 cm   3.7 mm

4 a Work out the circumference of the circle.

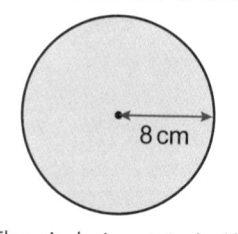
8 cm

The circle is cut in half.
  b Work out the length of the arc on the semicircle.

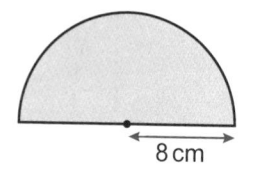
8 cm

  c Work out the total perimeter of the semicircle.

5 Work out the perimeter of this semicircle in terms of π.

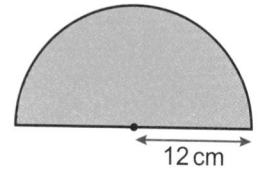
12 cm

6 Work out the perimeter of this quarter circle. Give your answer to 3 s.f.

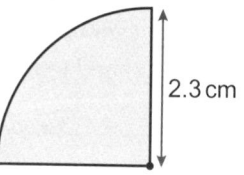
2.3 cm

7 A circular track has radius 10 m.

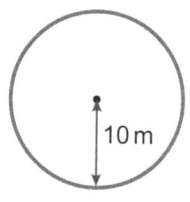
10 m

Work out, to the nearest m, how far you would run if you ran
  a round the whole track
  b three times round the track
  c round $\frac{1}{2}$ of the track
  d round $\frac{1}{4}$ of the track
  e round $\frac{1}{6}$ of the track.

8 **R** A cake has radius 10 cm.

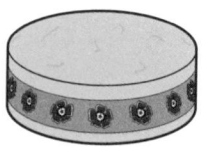

  a Work out the circumference of the cake in terms of π.
  b Diana cuts it into pieces with angles of 30°.
    i How many pieces can she cut?
    ii What fraction of the cake is each slice?

9 **R** Repeat **Q8** part **b** for these angle sizes.
  a 60°   b 40°   c 120°

10 Work out the arc length of each sector. Give your answers to an appropriate degree of accuracy.

Example

a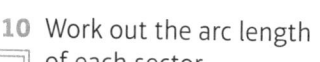
8 cm 20°

b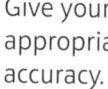
120°
12.2 cm

c
18.4 mm
270°

11 Work out the area of each sector. Give your answers correct to 3 s.f.

a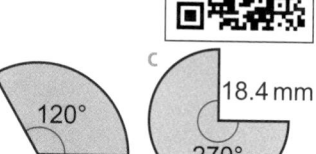
5.4 cm
48°

b
8.2 cm
63°

c
223°
11.4 mm

12 Work out the perimeter of each sector. Give your answers correct to 3 s.f.

a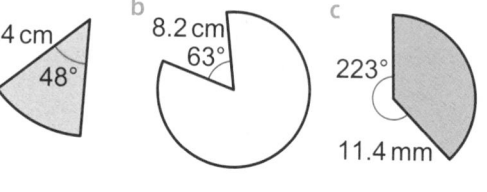
2 mm
72°

b
67°
18.1 cm

13 **Exam-style question**

ABC is a sector of a circle, centre B.

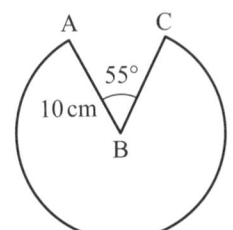

The radius of the circle is 10 cm.

The angle of the sector removed is 55°.

Calculate the area of the remaining shape.

Give your answer correct to 3 significant figures. **(3 marks)**

17.5 Composite 2D shapes and cylinders

1 Work out the perimeter and area of each of these shapes. Give your answers to an appropriate degree of accuracy.

Example

a

7 cm   2.25 cm   b

4.5 cm

60° 5.6 cm
60°   60°

2 A company logo is made of two quarter circles with radius 11.2 mm and a square with side length 11.2 mm.

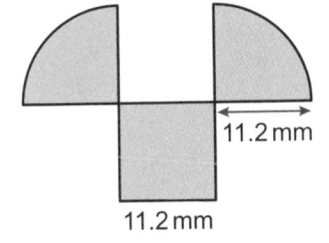

11.2 mm

11.2 mm

a Work out
  i the area of one of the quarter circles
  ii the perimeter of one of the quarter circles.
b Work out the area of the logo.
c Work out the perimeter of the logo.

3 The diagram shows a mirror surrounded by a square frame. Calculate the area of the frame.

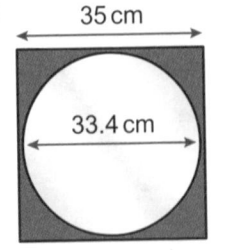

35 cm

33.4 cm

4 Which of these is the correct formula for the volume of a cylinder?

A $V = \pi r^2 h^2$

B $V = \pi r^2 h$

C $V = \pi r^3 h$

D $V = \pi r h^2$

E $V = \pi r^2$

5 Work out the volume of each cylinder. Give your answers to 3 s.f.

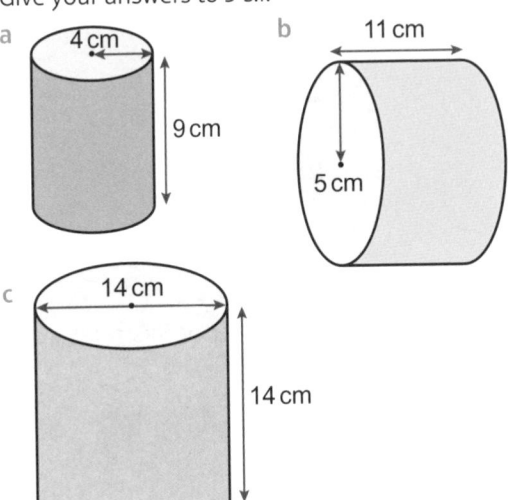

a   4 cm   b   11 cm

9 cm   5 cm

c   14 cm

14 cm

6 A soup can has a radius of 3 cm and a height of 11 cm.
  a What is the volume of the can?
  b How many cans would you need to hold 10 litres of soup?

Q6b hint 1 litre = 1000 cm³

7 **R** Which of these two paddling pools has the larger volume?

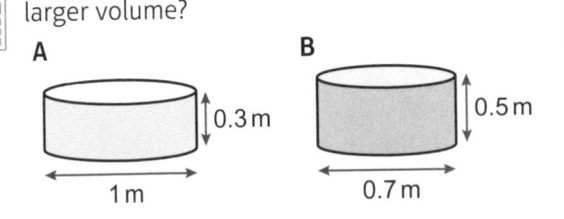

A   B

0.3 m   0.5 m

1 m   0.7 m

8

> **Exam-style question**
>
> A cylindrical hosepipe is 15 m long.
> The diameter of the circular end is 2 cm.
>
> What is the volume of the hosepipe?
>
> Give your answer in litres to 3 significant
> figures. **(4 marks)**

9 Petrol is held in a cylindrical tank.
The tank is 12 m long and has a radius of 2 m.
a Work out the volume of the tank.
The drum is completely filled with petrol.
Petrol has a density of 720 kg/m³.
b Work out the mass of the petrol in the tank
in kilograms.

> **Q9b hint** Density = $\dfrac{\text{mass}}{\text{volume}}$

10 Work out the total
surface area of each
cylinder in **Q5**.

Example

11 A drainpipe is 3 m long.
Its cross-section has diameter 15 cm.
Work out in terms of π
a the volume   b the surface area.

12 Work out the area of cardboard needed to
make the inside of a kitchen roll.

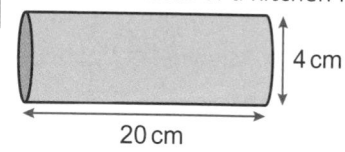
4 cm
20 cm

17.6 Pyramids and cones

1 Work out the volume of each pyramid.

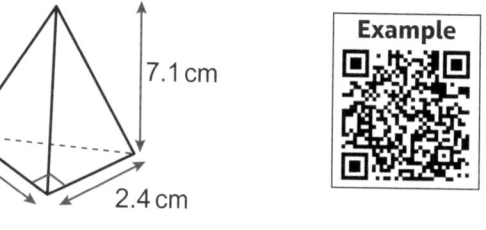
a
7.1 cm
3.5 cm
2.4 cm

Example

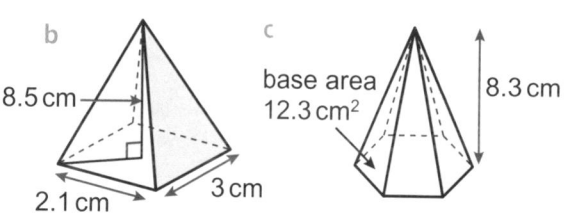
b
8.5 cm
2.1 cm
3 cm

c
base area
12.3 cm²
8.3 cm

2 A pyramid tea bag has a square base with
width 24 mm. It has height 32 mm.
Work out the volume of the tea bag in mm³.

3 Work out the surface area of
each pyramid.

Example

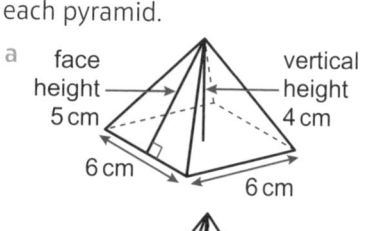
a
face
height
5 cm
vertical
height
4 cm
6 cm
6 cm

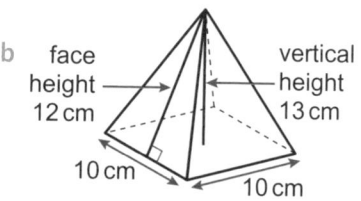
b
face
height
12 cm
vertical
height
13 cm
10 cm
10 cm

4 Work out the volume of each pyramid
in **Q3**.

5 By rounding the dimensions to 1 s.f. work out
an estimate of
a the volume of the pyramid
b the surface area of the pyramid.

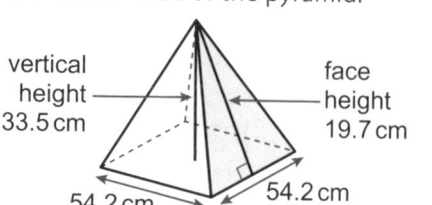
vertical
height
33.5 cm
face
height
19.7 cm
54.2 cm
54.2 cm

6 Which of these is the correct formula for the
volume of a cone?

A $V = \dfrac{\pi r^2 h}{2}$ **B** $V = \dfrac{\pi r^2}{3}$

C $V = \dfrac{\pi r^2 h}{3}$ **D** $V = \pi r^2 h$

7 Work out the volume of each cone.

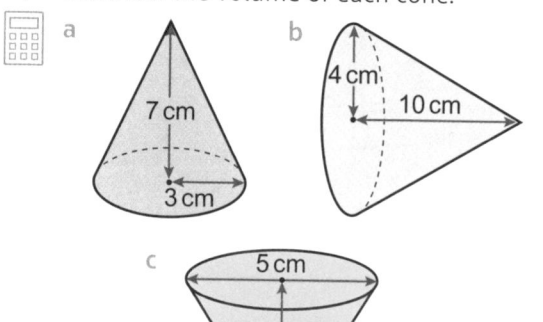
a
7 cm
3 cm
b
4 cm
10 cm
c
5 cm
5 cm

8 An ice-cream cone has a diameter of 6 cm and a height of 15 cm.

a What is the radius of the base of the cone?

b What is the volume of the cone? Give your answer correct to 3 s.f.

c How many of these cones can be completely filled from a 2 litre box of ice cream?

> **Q8c hint** 1 cm³ = 1 ml
> 1 litre = 1000 ml

9 Work out the total surface area of each cone

i in terms of π

ii as a number correct to 3 s.f.

Example

a

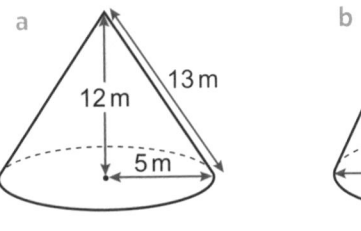
13 m
12 m
5 m

b

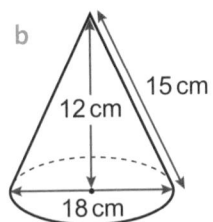
15 cm
12 cm
18 cm

c

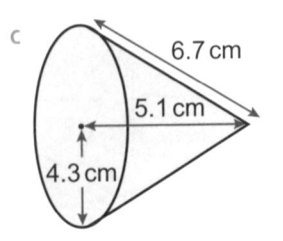
6.7 cm
5.1 cm
4.3 cm

10 Work out the volume of each cone in **Q9**

i in terms of π

ii as a number correct to 3 s.f.

11 Exam-style question

A conical water cup is shown.

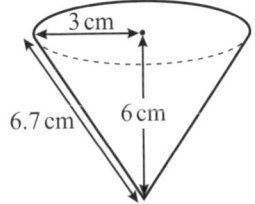
3 cm
6.7 cm
6 cm

Work out the surface area of the cup. **(3 marks)**

17.7 Spheres and composite solids

1 A tennis ball has radius 2.6 inches. Work out the volume of the tennis ball

a in terms of π

b to the nearest inch³.

Example

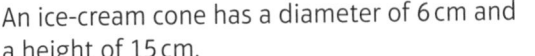

2 A football has a diameter of 25 cm. Work out the volume of the football to the nearest cm³.

> **Q2 hint** Work out the radius first.

3 **R** A hemisphere has a radius of 12 m. What is the volume?

A 144π m³

B 2304π m³

C 2304 m³

D 1152π m³

12 m

4 Work out the surface area of the football in **Q2**. Give your answer to the nearest cm³.

5 The Earth can be modelled as a sphere with a diameter of 12 742 km.
Work out an estimate of the surface area of the Earth.
Give your answer to 3 significant figures.

6 Work out the total surface area of the hemisphere in **Q3**. Give your answer

a in terms of π

b to 2 decimal places (2 d.p.)

Example

7 a Work out the volume of this solid.

b Work out the surface area of the solid.

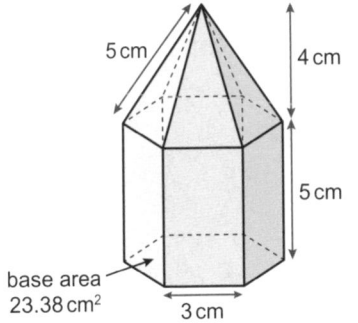
5 cm
4 cm
5 cm
base area 23.38 cm²
3 cm

8 Work out the volume of each solid. Give your answers in terms of π.

a

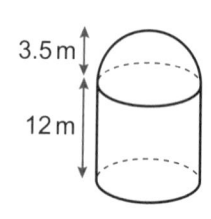

8.5 cm
6.5 cm
8.5 cm

b

7 cm
12 cm

c

3.5 m
12 m

9 a Use Pythagoras' theorem to work out the slant height of the cone.

8 cm
6 cm

Copy and complete the working.

$6^2 + 8^2 = \square$

Slant height, $l = \sqrt{\square} = \square$

b Work out the area of the curved surface in terms of π.

c Work out the area of the base in terms of π.

d Work out the total surface area of the cone in terms of π.

e Work out the total surface area of the cone as a decimal correct to 1 d.p.

10 Work out the total surface area of each solid
  i in terms of π
  ii as a decimal correct to 1 d.p.

a

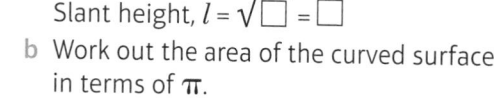

14 cm
9 cm

b

8.2 cm
9.4 cm

11 Work out the total surface area of each solid in **Q8**
  i in terms of π
  ii as a number correct to 3 s.f.

12

Exam-style question

A frustum is made by removing a small cone from a similar large cone.

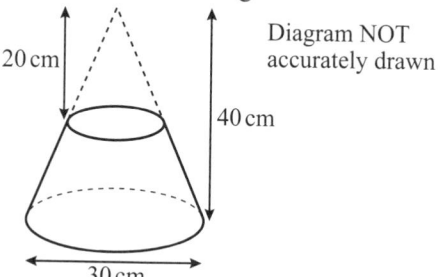

20 cm
40 cm
30 cm

Diagram NOT accurately drawn

The height of the small cone is 20 cm.

The height of the large cone is 40 cm.

The diameter of the base of the large cone is 30 cm.

Work out the volume of the frustum.

Give your answer correct to 3 significant figures. **(4 marks)**

March 2013, Q22, IMAO/2H

Q12 hint The smaller cone is similar to the larger cone. Since the height of the smaller cone is $\frac{1}{2}$ the height of the larger cone, the diameter will be $\square$ the width.

17 Problem-solving

Solve problems using these strategies where appropriate:

• **Use pictures**
• **Use smaller numbers**
• **Use bar models**
• **Use x for the unknown**
• **Use flow diagrams**
• **Use more bar models**
• **Use formulae**
• **Use arrow diagrams**
• **Use graphs.**

1 **R** Colin is worried that his 230 litre water butt is going to overflow. His roof collects 750 litres per 25 mm of rainfall. 10 mm of rain is expected.
Will the water butt overflow?

2 Jermaine and Teri took turns working at the school concert for a total of 2 hours. Jermaine worked for $\frac{5}{8}$ of the time and Teri worked the rest.
How many minutes did Teri work?

3 Felicia is buying a fringe to put round tables. Each table has a diameter of 145 cm.
a What length of fringe does she need for each table?
b Felicia sees that the fringe she wants is sold in 30 metre rolls.
How many whole tables can she trim with one roll?

4 In a triangle, angle ABC is 48° and angle ACB is twice the size of angle BAC.
What is the size of angle BAC?

5 Trevor is lining the outside edge of his vegetable patch with flexible plastic edging. The edging comes in rolls of 1.2 m long.

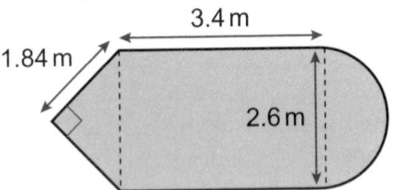

a How many rolls of edging will Trevor need to complete the perimeter?
b What is the area of Trevor's vegetable patch?
Write your answer to 2 d.p.

6 **R** Kirsty has a cylindrical plastic food container. The radius is 6.4 cm and the height is 12 cm.
Kirsty has 1500 ml of juice to put in the container.
Will the container hold all the juice?
(1 ml = 1 cm³)

7 Graham is fitting a semicircular stained glass window in his door.

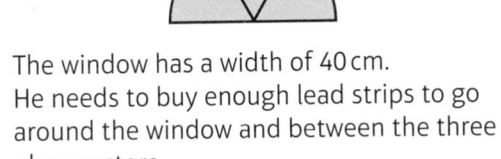

The window has a width of 40 cm.
He needs to buy enough lead strips to go around the window and between the three glass sectors.
How much lead will Graham need to buy?

8 The length of a rectangular garden plot is $n + 4$ and the width is $n + 2$.
Work out an expression for the area of the garden plot.
Simplify your answer.

9 **Exam-style question**

OAB is a sector of a circular garden, centre O.

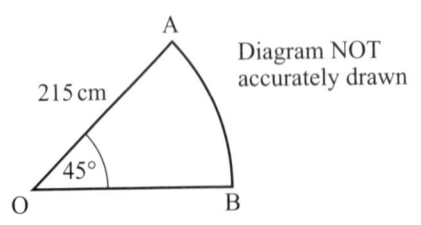

This sector is to be filled with red flowers.
The radius of the circle is 215 cm.
The angle of the sector is 45°.
Calculate the area of sector OAB.
Give your answer correct to 3 significant figures. **(2 marks)**

10 **R** Elias is doing a sand art project.
His container shape is a square-based pyramid on top of a cuboid.

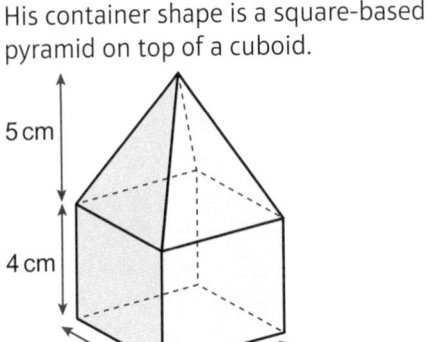

He has 100 cm³ of sand.
Does Elias have enough sand to fill his container?

Q10 hint Volume of a pyramid
$= \frac{1}{3} \times$ area of base $\times$ vertical height

18 FRACTIONS, INDICES AND STANDARD FORM

18.1 Multiplying and dividing fractions

1 Work out
a $1\frac{5}{8} \times \frac{2}{5} = \frac{\boxed{}}{8} \times \frac{2}{5} =$ b $2\frac{7}{10} \times \frac{3}{4}$
c $4\frac{1}{2} \times \frac{2}{3}$ d $\frac{1}{4} \times 5\frac{1}{3}$
e $\frac{3}{5} \times 1\frac{7}{8}$ f $\frac{5}{6} \times 2\frac{3}{4}$

2 Work out
a $3\frac{1}{3} \times \frac{2}{5}$ b $\frac{4}{5} \times 1\frac{3}{8}$
c $1\frac{1}{6} \times \frac{2}{3}$ d $3\frac{3}{4} \times \frac{3}{10}$
e $\frac{5}{6} \times 2\frac{1}{4}$ f $3\frac{3}{10} \times \frac{4}{11}$

3 Work out
a $1\frac{3}{4} \times 3\frac{1}{4} = \frac{\boxed{}}{4} \times \frac{\boxed{}}{4} =$ b $2\frac{1}{2} \times 4\frac{1}{2}$
c $3\frac{1}{3} \times 1\frac{5}{8}$ d $2\frac{1}{3} \times 2\frac{1}{5}$
e $1\frac{5}{7} \times 4\frac{1}{2}$ f $1\frac{5}{6} \times 2\frac{3}{7}$

4 Work out
a $1\frac{3}{4} \div \frac{1}{2}$ b $2\frac{1}{5} \div \frac{1}{3}$
c $1\frac{7}{8} \div \frac{4}{5}$ d $2\frac{1}{3} \div \frac{3}{7}$
e $\frac{7}{10} \div 1\frac{4}{5}$ f $\frac{2}{3} \div 1\frac{5}{6}$

Example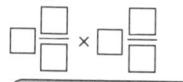

5 Work out
a $1\frac{3}{4} \div 2\frac{1}{10}$ b $3\frac{1}{4} \div 2\frac{1}{3}$
c $1\frac{1}{4} \div 2\frac{4}{5}$ d $1\frac{7}{8} \div 1\frac{5}{6}$
e $4\frac{1}{3} \div 2\frac{3}{4}$ f $1\frac{4}{5} \div 4\frac{7}{10}$

6 Use a calculator to work out
a $3\frac{4}{7} \times \frac{3}{5}$ b $11\frac{1}{2} \times 15\frac{3}{4}$
c $22 \div 3\frac{1}{5}$ d $16\frac{5}{7} \div 10\frac{1}{6}$

7 **P a** One of these cards doesn't have a pair.
Which one is it?

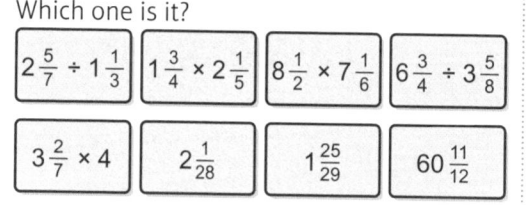

$2\frac{5}{7} \div 1\frac{1}{3}$ $1\frac{3}{4} \times 2\frac{1}{5}$ $8\frac{1}{2} \times 7\frac{1}{6}$ $6\frac{3}{4} \div 3\frac{5}{8}$

$3\frac{2}{7} \times 4$ $2\frac{1}{28}$ $1\frac{25}{29}$ $60\frac{11}{12}$

$3\frac{17}{20}$ $13\frac{1}{7}$ $6\frac{1}{3}$

b Make up a calculation that gives the answer that doesn't have a calculation card.

8 **P** A length of rope is cut into pieces measuring $1\frac{3}{4}$ m.
How many whole pieces can be cut from a 40 m rope?

9 **P** A window measures $1\frac{1}{4}$ m by $2\frac{1}{2}$ m.
What is the area of the window?

10 Each of these calculations is missing a single digit.
Work out the missing digits.
a $\frac{3}{5} \times 2\frac{\boxed{}}{4} = 1\frac{7}{20}$ b $\boxed{}\frac{1}{4} \times 2\frac{1}{2} = 8\frac{1}{8}$
c $\frac{5}{\boxed{}} \div 3\frac{1}{3} = \frac{1}{4}$ d $1\frac{3}{4} \div 2\frac{\boxed{}}{3} = \frac{3}{4}$

11 **P / R** Use each of the digits 3, 4, 5, 6, 7 and 8 once in the following calculation to give the largest solution result.

$$\frac{\boxed{}}{\boxed{}} \times \frac{\boxed{}}{\boxed{}}$$

12 **Exam-style question**

A bracelet is made of beads which are $1\frac{1}{3}$ cm long.

Each bead costs 20p and the thread costs £1.80.

The bracelet is $22\frac{2}{3}$ cm long.

What is the cost of making the bracelet?

(3 marks)

13 **P** There are 12 inches in a foot and approximately $2\frac{1}{2}$ centimetres in an inch.
Approximately how many centimetres are there in a foot?

18.2 The laws of indices

1 a Find the reciprocal of
i 0.1 ii 0.25 iii 0.05
iv 0.3 v −2 vi −0.5
b Multiply each number in part **a** by its reciprocal. What do you notice?

Q1a hint The **reciprocal** of a number is 1 divided by that number.

2 Copy and complete
   a $(2^5)^3 = 2^\square \times 2^\square \times 2^\square = 2^\square$
   b $(5^2)^5 = 5^\square \times 5^\square \times 5^\square \times 5^\square \times 5^\square = 5^\square$

3 Write as a single power
   a $(6^2)^4$
   b $(7^3)^5$
   c $(10^4)^3$
   d $(12^6)^4$

Example

4 The answer to each question is one of the following.
   4^2 4^3 4^5 4^6 4^7 4^8 4^9 4^{10}
   Choose the correct answer for each expression.
   a $(4^3)^3$ b $(4^2 \times 4^3)^2$
   c $(4^5 \div 4^3)^3$ d $(4^3)^2 \times 4$
   e $(4^{11} \div 4^8) \times 4^2$ f $(4^5 \div 4^4)^3$
   g $(4^2 \times 4)^3 \div 4$ h $\dfrac{4^3 \times 4^5}{(4^2)^3}$

5 **R** Copy and complete the pattern.
   $3^3 = 27$
   $3^2 = \square$
   $3^1 = \square$
   $3^0 = 1$
   $3^{-1} = \dfrac{1}{3^1} = \square$
   $3^{-2} = \dfrac{1}{3^2} = \square$
   $3^{-3} = \dfrac{1}{3^3} = \square$

   Use your calculator to work out
   a 5^0 b 80^0 c 24^0

7 Write as a fraction
   a 2^{-1} b 6^{-1} c 11^{-1} d y^{-1}

8 Work out the value of
   a 4^{-2} b 5^{-3} c 7^{-2}
   d $\left(\dfrac{1}{3}\right)^{-2}$ e $\left(\dfrac{1}{4}\right)^{-2}$ f $\left(\dfrac{3}{5}\right)^{-2}$

Example

9 Write as a single power.
   a $3^4 \times 3^{-1}$ b $(4^2)^{-2}$
   c $6^{-4} \times 6^4$ d $\dfrac{5^3 \times 5^{-1}}{5^2}$
   e $10^4 \div 10^{-5}$ f $(2^{-3})^2 \times 2^{-5}$

10 Write as a single power.
   a $x^{-3} \times x^7$ b $y^{-3} \times y^{-2}$
   c $\dfrac{k^{-1}}{k^3}$ d $\dfrac{a^4}{a^{-2}}$
   e $\dfrac{p^5 \times p^{-3}}{p^7}$ f $\dfrac{b^{-5} \times b^8}{b}$

11 Write these numbers from smallest to largest.
   10^5 100^3 10^{-5} 100^{-3} 1^6 1^{-6}

12 Work out
   a $5^{-1} \times 15$ b 120×10^{-1}
   c 2000×10^{-2} d 7×8^0

13 Use your calculator to work out these calculations.
   Where necessary, give your answers to 3 s.f.
   a $0.8^2 \times 10^{-1}$ b $7^9 \div 5^{10}$
   c $\dfrac{5^{-4} \times 6^3}{7^{-3}}$ d $4.8^3 \times 2.6^2 - 6.5^{-2}$

14 | Exam-style question
   a Find the value of 5^3. **(1 mark)**
   b Write down the reciprocal of 9. **(1 mark)**
   c Work out the value of $\left(\dfrac{3}{4}\right)^{-2}$. **(2 marks)**

15 | Exam-style question
   Write these numbers in order of size.
   Start with the smallest number.
   4^{-3}, 0.4, 0.4^2, 4^{-1} **(2 marks)**

16 | Exam-style question
   a Work out $2.7^{-3} + 11.5$.
     Write down all the figures on your calculator display. **(2 marks)**
   b Write your answer to part **a** to 3 significant figures. **(1 mark)**

18.3 Writing large numbers in standard form

1 Write each number as an ordinary number.
   The first one has been started for you.
   a $4 \times 10^3 = 4 \times 1000 = \square$
   b 9×10^2
   c 6×10^5
   d 8×10^8

2 Write each number in standard form.
   a 2000
   b 70 000
   c 300 000 000
   d 5 million

Example

3 Write each number as an ordinary number.
  a $5.6 \times 10^3 = 5.6 \times 1000 = \square$ b 7.4×10^5
  c 2.9×10^{10} d 1.75×10^4
  e 8.23×10^9 f 3.01×10^7
  g 4.621×10^5 h 9.92×10^3

4 Match the numbers on the left to those on the right.

  Example

  a 2800   **A** 2.83×10^2
  b 283 000   **B** 2.83×10^4
  c 28 000 000   **C** 2.8×10^6
  d 283   **D** 2.8×10^3
  e 2 800 000   **E** 2.83×10^8
  f 2 800 000 000   **F** 2.8×10^7
  g 283 000 000   **G** 2.83×10^5
  h 28 300   **H** 2.8×10^9

5 Write each number using standard form.
  a 75 000   b 240 000
  c 370 000   d 4900
  e 8 630 000   f 27 410 000
  g 903 000   h 325 million

6 Jason uses a calculator to work out
  94 500 × 370 000.
  His calculator gives the answer

  3.4965×10^{11}

  Write this number as an ordinary number.

7 **R** These numbers are not written correctly in standard form.
  Write them in standard form.
  a 44×10^4 b 31×10^6 c 12.5×10^2
  d 0.8×10^7 e 0.29×10^5 f 20.8×10^4

  Q7a hint $44 \times 10^4 = 4.4 \times 10 \times 10^4$

8 The table shows the populations of eight Asian countries in 2014 (rounded to the nearest million).

Country	Population
China	1 394 000 000
India	1 267 000 000
Indonesia	253 000 000
Pakistan	185 000 000
Philippines	101 000 000
Iran	78 000 000
Thailand	67 000 000
Singapore	6 000 000

Write each population in standard form.

9 A quadrillion is 10^{15}. Write 1 quadrillion
  a as an ordinary number
  b in standard form.

10 The table shows the meaning of prefixes for large numbers.

Prefix	Letter	Number
tera-	T	1 000 000 000 000
giga-	G	1 000 000 000
mega-	M	1 000 000
kilo-	k	1000

Write the following measurements in standard form.
  a 9 megagrams in grams
  b 6.2 GW in watts
  c 12.4 Tm in metres
  d 0.6 kilojoules in joules

11 **R** Write $<$ or $>$ in the boxes.
  a $2.6 \times 10^3 \ \square \ 9.2 \times 10^6$
  b $6.02 \times 10^8 \ \square \ 1.8 \times 10^5$
  c $1.5 \times 10^5 \ \square \ 8.3 \times 10^4$
  d $7.5 \times 10^6 \ \square \ 5.8 \times 10^6$
  e $1.09 \times 10^2 \ \square \ 1.9 \times 10^2$
  f $5.08 \times 10^4 \ \square \ 8.05 \times 10^4$

12 **Exam-style question**

  Write these numbers in order of size.
  Start with the largest number.
  6.4×10^5, 64 000, 640, 6.4×10^8, 6.4×10^3
  (2 marks)

13 **Exam-style question**

  a Write 20 900 000 in standard form.
  (1 mark)

  b Write 9.24×10^5 as an ordinary number.
  (1 mark)

18.4 Writing small numbers in standard form

1 Write each number as an ordinary number.
  The first one has been started for you.
  a $4 \times 10^{-3} = 4 \times 0.001 = \square$ b 7×10^{-4}
  c 2×10^{-1} d 6×10^{-7}

2 Write each number in standard form.

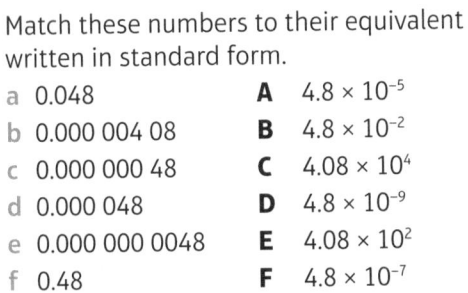

Example

a 0.009
b 0.1
c 0.06
d 0.000 008

3 Write each number as an ordinary number.

a $2.2 \times 10^{-4} = 2.2 \times 0.0001 = \square$
b 9.3×10^{-3}
c 5.29×10^{-6}
d 6.05×10^{-2}

4 Write each number in standard form.

Example

a 0.26
b 0.0038
c 0.000 21
d 0.000 987
e 0.0359
f 0.000 505
g 0.002 604
h 0.5253

5 Match these numbers to their equivalent written in standard form.

a 0.048   **A** 4.8×10^{-5}
b 0.000 004 08   **B** 4.8×10^{-2}
c 0.000 000 48   **C** 4.08×10^{4}
d 0.000 048   **D** 4.8×10^{-9}
e 0.000 000 0048   **E** 4.08×10^{2}
f 0.48   **F** 4.8×10^{-7}
g 40 800   **G** 4.08×10^{-6}
h 408   **H** 4.8×10^{-1}

6 **R** A question in an exam reads
Write 0.000 0503 in standard form.
Here are some of the incorrect answers given.
Oliver: 0.503×10^{-4}
Jade: 5.03×10^{-4}
Reece: 5.03×10^{5}

a For each answer explain what the student did wrong.
b What is the correct answer?

7 The table shows the meaning of prefixes for small numbers.

Prefix	Letter	Number
deci-	d	0.1
centi-	c	0.01
milli-	m	0.001
micro-	μ	0.000 001
nano-	n	0.000 000 001
pico-	p	0.000 000 000 001

Write the following measurements in standard form.

a 6 decilitres in litres
b 2.5 μm in metres
c 78 ps in seconds
d 12.5 nanojoules in joules
e 26.8 cm in metres
f 0.4 mg in grams

8 The table gives typical wavelengths of different types of electromagnetic waves.
a Write the wavelengths in standard form.

Type of wave	Wavelength (m)
radio wave	166.2
microwave	0.122
infrared	0.000 000 925
visible light	0.000 000 59
ultraviolet	0.000 000 25
X-ray	0.000 000 005
gamma ray	0.000 000 000 01

9 **Exam-style question**

a Write 0.000 002 63 in standard form. **(1 mark)**
b Write 7.01×10^{-3} as an ordinary number. **(1 mark)**

18.5 Calculating with standard form

1 Work out these calculations.
Give your answers in standard form.
a $2 \times 4 \times 10^{6} = \square \times 10^{6}$
b $3 \times 2.5 \times 10^{-4}$ c $5 \times 8 \times 10^{2}$ d $6 \times 4 \times 10^{-5}$
e $6 \times 10^{-4} \div 3 = \square \times 10^{-4}$
f $7 \times 10^{5} \div 2$ g $2 \times 10^{7} \div 4$ h $1 \times 10^{-6} \div 10$

2 **P** A grain of sand weighs 5×10^{-6} kg.
How many grains of sand are there in 1 kg?

3 Giving your answers in standard form, work out

Example

a $2 \times 10^{5} \times 3 \times 10^{4}$
b $6 \times 10^{2} \times 5 \times 10^{5}$
c $4 \times 10^{8} \times 7 \times 10^{-3}$
d $1.2 \times 10^{5} \times 5 \times 10^{-2}$
e $4 \times 10^{10} \times 2.2 \times 10^{-4}$
f $8 \times 10^{-2} \times 3.1 \times 10^{-3}$

4 Giving your answers in standard form, work out

a $\dfrac{6 \times 10^8}{2 \times 10^2}$ b $\dfrac{9 \times 10^{-7}}{3 \times 10^{-4}}$ c $\dfrac{3 \times 10^9}{6 \times 10^5}$

d $6.4 \times 10^4 \div 2 \times 10^{-2}$

e $7.59 \times 10^5 \div 1 \times 10^2$

f $3.5 \times 10^{-3} \div 5 \times 10^{-6}$

5 Use a calculator to work out these calculations.

Give your answers to 3 s.f.

a $3.6 \times 10^4 \times 6.1 \times 10^5$

b $2.9 \times 10^8 \times 8.25 \times 10^{-3}$

c $4.8 \times 10^{-2} \times 7.2 \times 10^{-4}$

d $(6.3 \times 10^2) \div (4.4 \times 10^7)$

e $(1.8 \times 10^5) \div (8.51 \times 10^{-3})$

f $\dfrac{2.65 \times 10^{-5}}{5.9 \times 10^6}$

> **Q5 hint** To round a standard form number to 3 significant figures, only round the number part.

6 **P** How many

a MW in 1 GW

b nm in 1 mm

c mg in 1 Mg?

7 **R / P** Bamboo grows at a rate of 3×10^{-3} m per hour.

How much will it grow in one week?

8 **Exam-style question**

Work out

a $7 \times 10^4 \times 3 \times 10^8$ **(2 marks)**

b $\dfrac{2 \times 10^5}{5 \times 10^{-2}}$ **(2 marks)**

Give your answers in standard form.

9 a Work out

i $4 \times 10^7 + 5 \times 10^6$

ii $9 \times 10^5 + 4 \times 10^3$

iii $2.5 \times 10^{-4} + 7.3 \times 10^{-3}$

iv $6.1 \times 10^{-4} + 3.8 \times 10^{-6}$

v $1.05 \times 10^6 + 2.4 \times 10^2$

vi $4.15 \times 10^{18} + 4.15 \times 10^{17}$

b Check your answers to part **a** using a calculator.

Example

10 a Work out

i $5 \times 10^7 - 6 \times 10^6$

ii $3 \times 10^{-4} - 9 \times 10^{-5}$

iii $2.8 \times 10^8 - 7.5 \times 10^6$

iv $4.7 \times 10^{-2} - 1.3 \times 10^{-4}$

v $7.4 \times 10^{-5} - 5.1 \times 10^{-7}$

vi $6.66 \times 10^{12} - 7.7 \times 10^9$

b Check your answers to part **a** using a calculator.

11 Use a calculator to work out

a $5.3 \times 10^8 - 9.2 \times 10^5$

b $8.4 \times 10^{-6} + 3.59 \times 10^{-10}$

c $1.7 \times 10^7 + 6.231 \times 10^9 - 9.42 \times 10^8$

12 Use a calculator to work out these calculations.

Give your answers to 3 significant figures.

a $\dfrac{4.1 \times 10^5 + 6.9 \times 10^3}{7.5 \times 10^2}$

b $6.3 \times 10^7 \times 5.2 \times 10^{-3} - 8.5 \times 10^4$

c $\dfrac{3.9 \times 10^4}{8.1 \times 10^8 + 4.7 \times 10^{-4}}$

13 **P** The mass of one iron atom is 9.28×10^{-23} g. How many atoms are there in one kilogram of iron?

14 The distance between Earth and the Sun is about 150 million km.

a Write this distance in metres.

b Write your distance from part **a** in standard form.

c It takes 500 s for light to travel from the Sun to Earth.

Use the formula speed $= \dfrac{\text{distance}}{\text{time}}$ to calculate an estimate for the speed of light.

15 **P** The closest distance from Earth to Jupiter is 5.88×10^8 km.

The highest speed a manned spacecraft ever achieved was 11.08 km/s, in 1969.

How long would it take a spacecraft travelling at this speed to reach Jupiter?

Give your answer in days to 3 significant figures.

18 Problem-solving

Solve problems using these strategies where appropriate:

- Use pictures
- Use bar models
- Use flow diagrams
- Use formulae
- Use graphs.
- Use smaller numbers
- Use x for the unknown
- Use more bar models
- Use arrow diagrams

1 **R** Joe has a bag of 7 balls numbered 1 to 7.
Joe wants to find the number 6.
He picks out a ball.

a What is the probability of not picking the number 6 ball?

Joe picks out the number 5.
He places it on the table and tries again to find the number 6 ball.

b Is it more or less probable that the number 6 ball will be picked out now than before? Explain how you know.

2 **R** Paddy is drawing a scale drawing of her new room so that she can see where to fit all her furniture.
The room is 345 cm × 420 cm.
Paddy uses a scale of 1 : 15.

a What size does she draw the room?

Her bed measures 96 cm × 198 cm.

b What size does she draw the bed?

3 A carpet tile measures $11\frac{5}{8}$ inches × $8\frac{4}{5}$ inches.

a What is the area of one tile?

b How many square inches will 25 tiles cover?

4 **R** Ben is finding a postal tube to send a tube of biscuits to his cousin at college.
The postal tube has a circumference of 25 cm rounded to the nearest centimetre.
The tube of biscuits has a diameter of 8 cm rounded to the nearest centimetre.
Explain why Ben will have to check that the tube of biscuits fits before he buys the postal tube.

5

Exam-style question

Work out the value of $\dfrac{6^3 \times 6^2}{6^4}$

Give your answer as a power of 6.

(2 marks)

6 David is looking at the export figures (£) from the UK for 2014.

Category	Value (£)
cereals and bakery	1 034 000 000
meat	865 000 000
dairy	780 000 000
fish and seafood	743 000 000
tea, coffee and cocoa	468 000 000

He gets confused by so many zeros.

a What can David do to make it easier to read the numbers?

b Write each of the figures in standard form.

7 Order these 2014 population figures from least to greatest.

Country	Population
Pakistan	1.85×10^8
British Virgin Islands	29 000
Greenland	5.7×10^4
Thailand	6.7×10^7
South Africa	5.3×10^7
USA	3.23×10^8
Belarus	9.3×10^6
Fiji	887 000

8 **R** Light takes 0.000 005 seconds to travel one mile.

a Write this time in standard form.

b Dori reads that it takes light one nanosecond to travel one foot.
Knowing that a nanosecond is one billionth of a second, how can Dori write this time in seconds in standard form?

9 An average adult human has 5.2×10^6 red blood cells in one mm^3 of blood. When people donate blood, about 500 ml is usually taken.
This is equal to 500 000 mm^3.
Approximately how many red blood cells are withdrawn?

10 Distance in space is often measured in astronomical units (AU).
1 AU measures approximately 9.3×10^7 miles.
This is approximately the distance from Earth to the Sun

a Neptune is approximately 30.1 AU from the Sun.
How many miles is this distance?

b The Oort Cloud can be as far as 1×10^5 AU from the Sun.
How many miles is this?

19 CONGRUENCE, SIMILARITY AND VECTORS

19.1 Similarity and enlargement

1 | Exam-style question |

A small triangle has base length 8 cm and height 3 cm.

Mirka enlarges the triangle.

The large triangle has base length 24 cm.

Diagram NOT accurately drawn

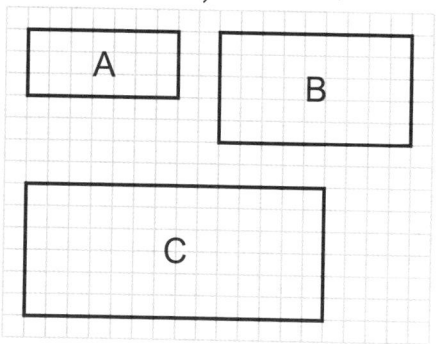

3 cm

8 cm   24 cm

Small triangle   Large triangle

The two triangles are similar.

Work out the height of the larger triangle.

(3 marks)

2 A shape is enlarged by scale factor 3.
What can you say about the shape and its enlargement?

3 Identify the two similar rectangles.
Give a reason for your answer.

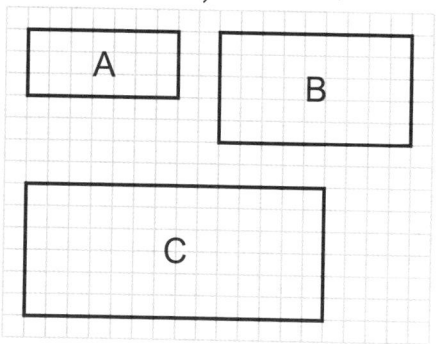

4 These two triangles are similar.

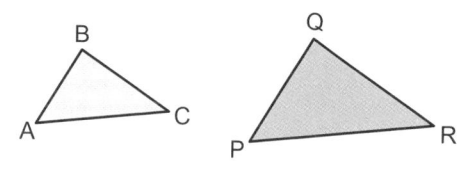

Measure all the angles in the triangles.
What do you notice?

5 Quadrilateral ABCD is similar to quadrilateral LMNO.

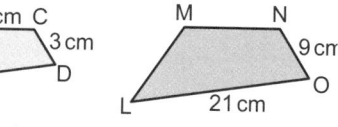

B 5 cm C
3 cm
A   D

M   N
9 cm
L   21 cm   O

a Which side corresponds to
  i AD   **ii** LM?
b What is the scale factor of the enlargement that maps
  i shape ABCD to shape LMNO
  ii shape LMNO to shape ABCD?
c Work out the length of
  i MN   **ii** AD
d Copy and complete.
  i NO = 3 × ☐   **ii** MN = 3 × ☐
  iii LM = ☐ × AB   **iv** LO = ☐ × ☐
e Which angle is the same as
  i angle BAD   **ii** angle LMN
  iii angle ADC?
f Write $\frac{CD}{NO}$ as a fraction.

Then write $\frac{AD}{LO}$ as a fraction.
What do you notice?

6 Triangles ABC and XYZ are similar.
AC and XZ are corresponding sides.

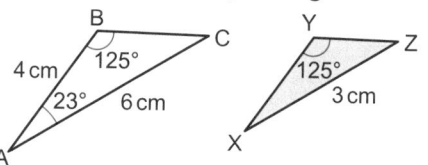

B
4 cm 125°   C
23°   6 cm
A

Y
125°   Z
3 cm
X

a Which side corresponds to BC?
b Work out the length of XY.
c Work out the size of
  i angle YXZ   **ii** angle YZX.

7 a Are triangles ABC and PQR similar?
Explain your answer.

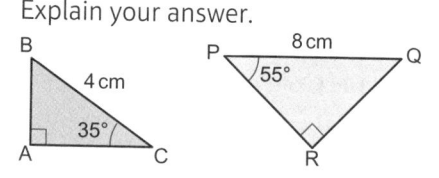

B
4 cm
35°
A   C

P   8 cm   Q
55°
R

b Which side corresponds to AB?
c Copy and complete.
  i PR = ☐ × AB **ii** PQ = 2 × ☐ **iii** QR = ☐ × ☐

| Q7a hint Work out all the angles. |

19.2 More similarity

1 **Exam-style question**

Triangles ABC and DEF are mathematically similar.

Angle A = angle D
Angle B = angle E
Angle C = angle F

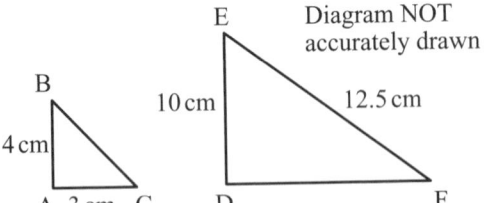

Diagram NOT accurately drawn

a Work out the length of DF. **(2 marks)**
b Work out the length of BC. **(2 marks)**

2 **a** Explain how you know that these two shapes are similar.

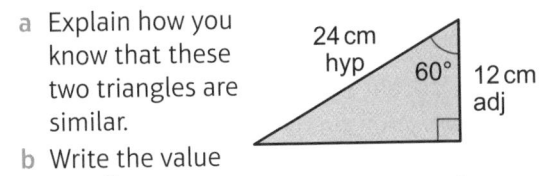

b Write the scale factor of the enlargement that maps shape A to shape B.
c Use the scale factor to find the value of x.

3 **a** Use the diagram in **Q2** to copy and complete
$$\frac{x}{8} = \frac{6}{\square}$$
b Solve the equation in part **a** to find the value of x.

4 **a** Explain how you know that these two triangles are similar.
b Write the value of $\frac{adj}{hyp}$
c The ratio $\frac{adj}{hyp}$ is called the cosine of the angle.
Check that $\cos 60° = \frac{1}{2}$

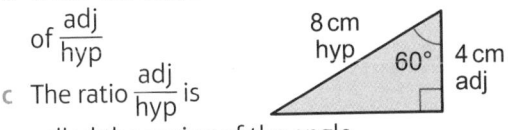

Q4b hint The value of $\frac{adj}{hyp}$ is always the same when the angle is 60° because the triangles are similar.

5 Find the values of x and y.
a

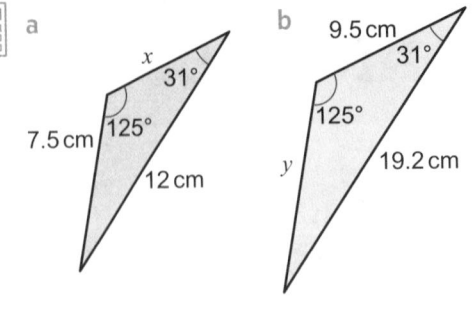

b

6 **R** The lines AC and PQ are parallel.

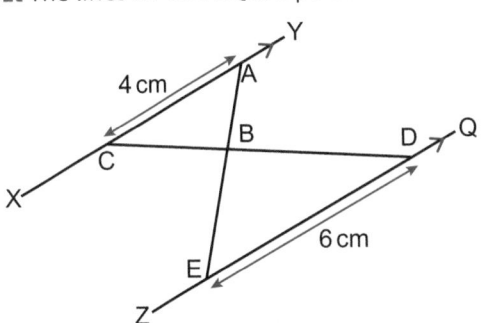

Example

Triangles PBQ and ABC are similar.
a Find the size of angle ACB.
b Which side corresponds to PQ?
c Write the scale factor of the enlargement that maps triangle PBQ to triangle ABC.
d Work out the length of QB.

7 **R** The lines XY and ZQ are parallel.

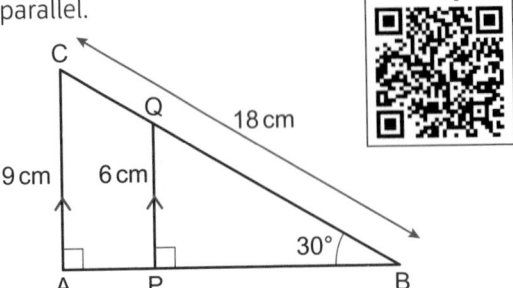

a Match each angle from column A to an equal angle in column B.

A	B
∠BAC	∠DEB
∠ABC	∠EDB
∠ACB	∠DBE

b Explain why triangles ABC and BDE are similar.
c CD = 7.5 cm
Work out the length of
i BC   **ii** BD

8 **R** The lines AB and CE are parallel. Triangles ABD and ECD are similar.

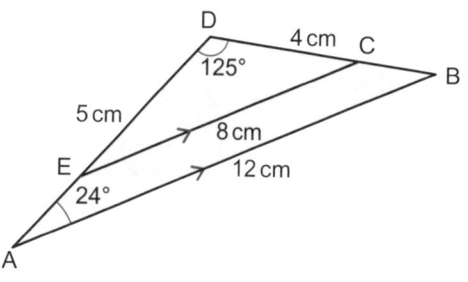

a Work out the size of angle DCE.
b Write the scale factor of the enlargement that maps triangle ABD to triangle ECD.
c Work out the length of AD.
d Work out the length of BC.

9 **P** a Show that triangle ABE and triangle ACD have the same angles, so they are similar.
b Work out the scale factor of the enlargement that maps triangle ABE to triangle ACD.
c Work out the length of CD.

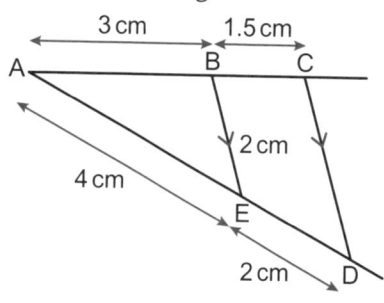

Q9a hint What do you know about the angles made when you have a pair of parallel lines?

19.3 Using similarity

1 These quadrilaterals have the same angles. Are they similar? Explain your answer.

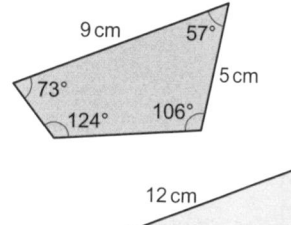

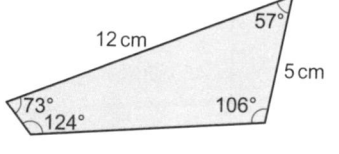

2 These pentagons have sides in the same ratio. Are they similar? Explain your answer.

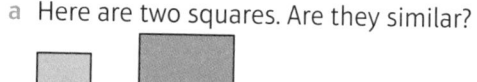
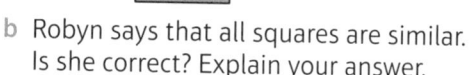

3 For each statement, write 'True' or 'False'.
a If two pentagons share 5 identical angles, they are always similar.
b If two triangles share 2 identical angles, they are always similar.
c If two polygons share the same angles, they are always similar.
d If two polygons share the same angles and have sides in the same ratio, they are always similar.
e If two triangles have three pairs of sides in the same ratio, they are always similar.

4 a Here are two squares. Are they similar?

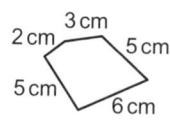

b Robyn says that all squares are similar. Is she correct? Explain your answer.

5 a Explain why two regular pentagons are similar.
b Explain why any two of the same type of regular polygon are similar.

6 **R** Two circles with different diameters are drawn. Explain why they are similar.

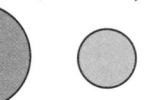

7 **P** In the diagram, shape A is mapped to shape B by an enlargement.

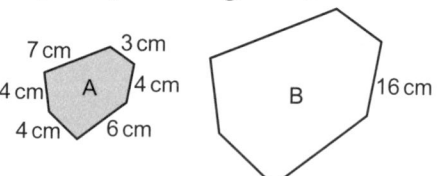

a Find the scale factor of the enlargement.
b Work out the lengths of the unknown sides of shape B.
c Work out the perimeter of shape A.
d Work out the perimeter of shape B.
e Copy and complete.
Perimeter of shape B = ☐ × perimeter of shape A
f Look at your answers to parts **a** and **e**. What do you notice?

8 **P** These two shapes are mathematically similar.

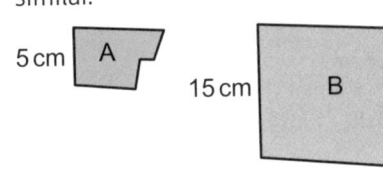

5 cm — A
15 cm — B

Shape A has perimeter 23 cm.
Work out the perimeter of shape B.

> **Q8 hint** Work out the scale factor of the enlargement using the two corresponding sides given.

9

Exam-style question

The diagram shows a company's logo.
It uses two similar shapes.

1 cm
4 cm

The perimeter of the smaller arrow is 7 cm.
Work out the perimeter of the logo.

(2 marks)

19.4 Congruence 1

1 Here are two pairs of congruent triangles.
Work out the size of angles x and y.

a

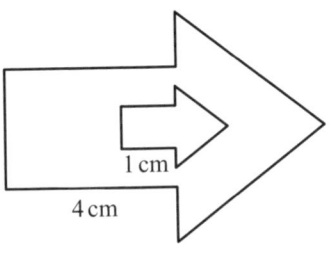

5 cm
55° 65°
4 cm

4 cm
y
5 cm
x

b

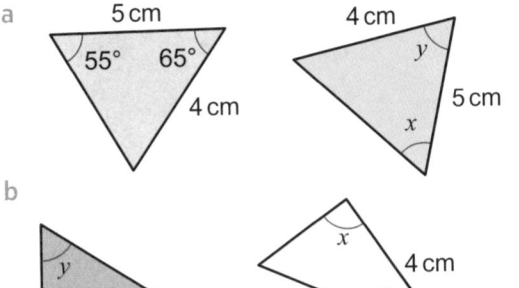

y
4 cm

x
4 cm
32°

2 Construct these triangles accurately using a ruler and compasses.
Are your two triangles congruent?

6 cm — B
A
7 cm
9 cm
C

7 cm — Y
Z
9 cm
6 cm
X

3 ABCDEF is a regular hexagon.

Example

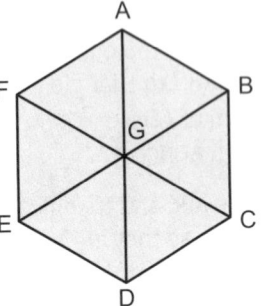

A
F   B
G
E   C
D

a What can you say about triangles ABG and BCG?

b Which angles are the same as angle GAB?

c Are triangles ABG and BCG congruent?
   If so, give a reason for congruency (SSS, SAS, etc.).

4 **P** ABC is a triangle.

C
38°
A   3 cm   X   3 cm   B

a Find the size of angle ABC.
b Find the size of angle BCX.
c Are triangles ACX and BCX congruent?
   If so, give a reason for congruency.

> **Q4 hint** What can you say about triangles ACX and BCX?

5 **P** The diagram shows a rectangle ABCD.

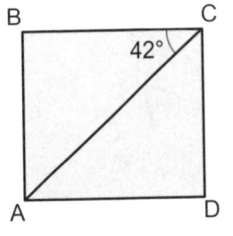

B   C
42°
A   D

Find the size of
a angle CAD   b angle BAC.

6 **P** The diagram shows a rhombus ABCD.
AC bisects angles BAD and BCD.

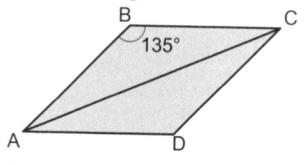

a Are triangles ABC and ACD congruent?

b Which angles are the same as angle BAC?

7 **R** List three pieces of information you would need in order to check whether two triangles are congruent.

> **Q7 hint** Think about the minimum information you need in order to accurately construct a unique triangle.

8 **P** Four points A, B, C and D are marked on the circumference of a circle.
They are joined to the centre of the circle E.

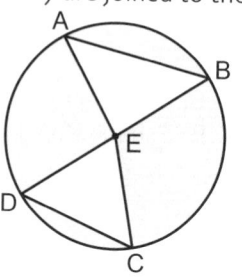

a Which lines are equal in length to AE? Give a reason for your answer.

b Which angles are equal to
  i angle ABE   ii angle EDC?
  Give a reason for your answer.

c Are triangles ABE and CDE congruent? Give a reason for your answer.

9

Exam-style question

The spokes of a cartwheel are set at regular intervals around the circumference and pass through the centre of the wheel at G.

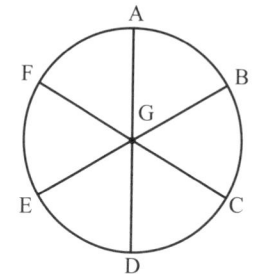

a Are shapes ABG and BCG congruent?
(2 marks)

b Work out the size of angle AGE.
(1 mark)

19.5 Congruence 2

1 Which triangle is congruent to triangle A?

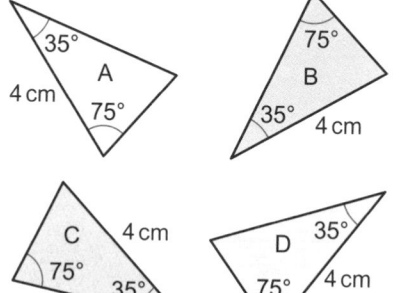

2 **R** ABCD is a rectangle.
M is the midpoint of AC and also of BD.

Example

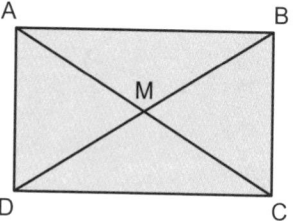

a Copy the diagram.
  Label any sides or angles that are equal.

b Are triangles ABM and CDM congruent?

3 **R** In the diagram, M is the midpoint of AC and of BD.

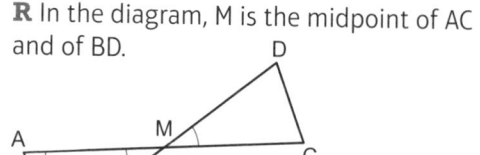

a Are triangles ABM and DCM congruent? Explain your answer.

b Work out the size of angle MCD.

c Work out the size of angle MDC.

d AM is 4 cm and BM is 3 cm.
  What length is DM?

4 **R** M is the midpoint of BD.

a Work out the size of angle MDC.

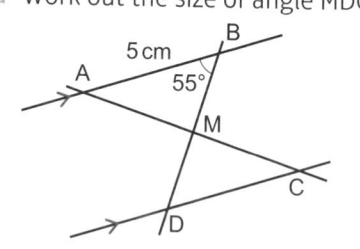

b Are triangles ABM and CDM congruent? Give a reason for your answer.

c AB = 5 cm. Which other side has length 5 cm?

5 **R** Draw these pairs of right-angled triangles accurately using a ruler, protractor and compasses.
Are the triangles in each pair congruent? Give a reason for your answer.

a
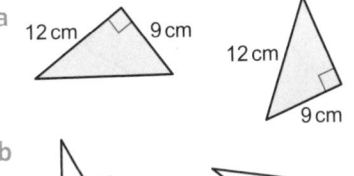

b
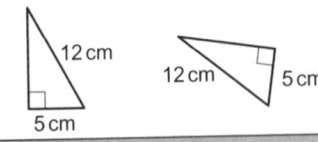

6
M is the centre of regular pentagon ABCDE.

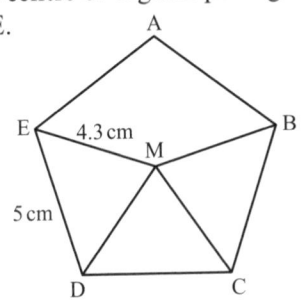

a Find the length of MD. **(1 mark)**
b Find the length of MC. **(1 mark)**
c Give three different reasons why triangles EMD and DMC are congruent. **(3 marks)**

7 **R** ABCD is a rectangle. M is the midpoint of AC and of BD.

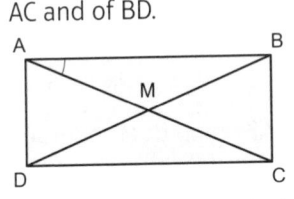

a Which angle in triangle ACD is equal to angle BAM?
b Give one length that is equal to AM.
c Are triangles ABC and ACD congruent? Give a reason for your answer.

19.6 Vectors 1

1 Describe each of these translations as column vectors.

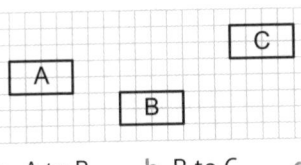

a A to B   b B to C   c C to A

2 Add these column vectors. Give your answers in the form of a single vector $\begin{pmatrix} \square \\ \square \end{pmatrix}$

a $\begin{pmatrix} 3 \\ 5 \end{pmatrix} + \begin{pmatrix} 2 \\ 1 \end{pmatrix} = \begin{pmatrix} \ \\ \ \end{pmatrix}$ b $\begin{pmatrix} 2 \\ 7 \end{pmatrix} + \begin{pmatrix} 5 \\ 4 \end{pmatrix}$

c $\begin{pmatrix} 0 \\ 5 \end{pmatrix} + \begin{pmatrix} 3 \\ 1 \end{pmatrix}$ d $\begin{pmatrix} 2 \\ -1 \end{pmatrix} + \begin{pmatrix} -3 \\ 4 \end{pmatrix}$

e $\begin{pmatrix} -1 \\ 4 \end{pmatrix} + \begin{pmatrix} -2 \\ 5 \end{pmatrix}$ f $\begin{pmatrix} 7 \\ -7 \end{pmatrix} + \begin{pmatrix} -2 \\ -1 \end{pmatrix}$

3 Shape X is translated to shape Y by the column vector $\begin{pmatrix} 3 \\ 4 \end{pmatrix}$.
Shape Y is translated to shape Z by the column vector $\begin{pmatrix} -2 \\ 1 \end{pmatrix}$.
Find the column vector of the single translation from shape X to shape Z.

4 Shape P is translated to shape Q by the column vector $\begin{pmatrix} 7 \\ -2 \end{pmatrix}$.
Shape Q is translated to shape R by the column vector $\begin{pmatrix} 5 \\ -1 \end{pmatrix}$.
Find the column vector of the single translation from shape P to shape R.

5 Write as column vectors
a $\overrightarrow{AB}$ b $\overrightarrow{CD}$ c $\overrightarrow{EF}$

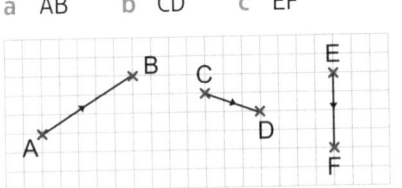

6 a Write as column vectors
i $\overrightarrow{PQ}$ ii $\overrightarrow{QR}$ iii $\overrightarrow{PR}$

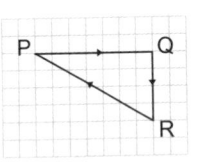

b Show that $\overrightarrow{PQ} + \overrightarrow{QR} = \overrightarrow{PR}$.

7 Copy and complete.
a $\overrightarrow{AB} + \overrightarrow{BC} = \square$
b $\overrightarrow{PQ} + \square = \overrightarrow{PT}$
c $\square + \overrightarrow{NA} = \overrightarrow{MA}$

8 Find the resultant of
a $\begin{pmatrix} -2 \\ 4 \end{pmatrix}$ and $\begin{pmatrix} 6 \\ 1 \end{pmatrix}$ b $\begin{pmatrix} 2 \\ -4 \end{pmatrix}$ and $\begin{pmatrix} -3 \\ 3 \end{pmatrix}$

c $\begin{pmatrix} 3 \\ -2 \end{pmatrix}$ and $\begin{pmatrix} 1 \\ -2 \end{pmatrix}$

9 Here are two vectors, **m** and **n**.

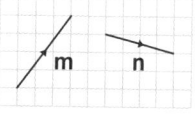

a Copy vector **m**.

b Draw vector **n** at the end of vector **m**.

c Draw in the third side of the triangle and label it **m** + **n**.

d Repeat the steps in parts **a**–**c**, but this time draw **n** first and draw **m** at the end of **n**. Label it **n** + **m**.

e Does the order in which you add vectors matter?

10 $\mathbf{m} = \begin{pmatrix} 3 \\ 2 \end{pmatrix}$ $\mathbf{n} = \begin{pmatrix} -1 \\ 2 \end{pmatrix}$ $\mathbf{q} = \begin{pmatrix} 3 \\ -2 \end{pmatrix}$

Copy the diagram and label the vectors using **m, n, q, −m, −n, −q, m + n, −q + n**

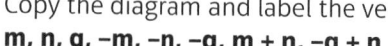

19.7 Vectors 2

1 Copy the diagram and label all of the vectors.

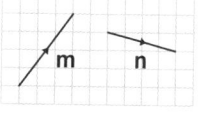

2 Subtract these column vectors.

a $\begin{pmatrix} 3 \\ 4 \end{pmatrix} - \begin{pmatrix} 1 \\ 5 \end{pmatrix}$ b $\begin{pmatrix} -2 \\ 7 \end{pmatrix} - \begin{pmatrix} 1 \\ 4 \end{pmatrix}$

c $\begin{pmatrix} 10 \\ 5 \end{pmatrix} - \begin{pmatrix} -3 \\ -5 \end{pmatrix}$

3 $\mathbf{x} = \begin{pmatrix} 3 \\ 5 \end{pmatrix}$ $\mathbf{y} = \begin{pmatrix} -2 \\ 1 \end{pmatrix}$ $\mathbf{z} = \mathbf{x} - \mathbf{y}$

a Write **z** as a column vector.

b Show **x**, −**y**, and **z** as a triangle on a grid.

4 $\mathbf{a} = \begin{pmatrix} -2 \\ -1 \end{pmatrix}$ $\mathbf{b} = \begin{pmatrix} 0 \\ -1 \end{pmatrix}$ $\mathbf{c} = \mathbf{a} - \mathbf{b}$

a Write **c** as a column vector.

b Show **a**, −**b** and **c** as a triangle on a grid.

5 $\mathbf{a} = \begin{pmatrix} -4 \\ 2 \end{pmatrix}$ $\mathbf{b} = \begin{pmatrix} 5 \\ 3 \end{pmatrix}$

Write as column vectors

a 3**a**   b 2**b**   c −2**a**   d −5**b**

6 $\mathbf{m} = \begin{pmatrix} -1 \\ 1 \end{pmatrix}$ $\mathbf{n} = \begin{pmatrix} 3 \\ 0 \end{pmatrix}$

Write as column vectors

a 3**m**   b −2**n**   c 2**n** + **m**   d 3**n** − **m**

7 $\mathbf{s} = \begin{pmatrix} 2 \\ -3 \end{pmatrix}$

Show on a grid

a **s**   b 3**s**   c −2**s**

8 $\mathbf{t} = \begin{pmatrix} 3 \\ -1 \end{pmatrix}$

Write as column vectors

a a vector in the same direction as **t** but three times as long

b a vector the same length as **t** but in the opposite direction

c a vector half as long as **t** but in the opposite direction

d a vector parallel to **t** but twice as long.

> **Q8d hint** A parallel vector may have the same direction or the opposite direction. There are two possible answers.

9 $\mathbf{w} = \begin{pmatrix} 2 \\ -5 \end{pmatrix}$

Choose from the vectors in the box to find a vector that is

$\begin{pmatrix} -4 \\ 10 \end{pmatrix}$ $\begin{pmatrix} -10 \\ 25 \end{pmatrix}$

$\begin{pmatrix} -6 \\ 15 \end{pmatrix}$ $\begin{pmatrix} 6 \\ -15 \end{pmatrix}$

a three times as long as **w**

b in the opposite direction to **w**

c twice as long as **w**

d parallel to **w** but 5 times as long

You can only choose each vector once.

19 Problem-solving

Solve problems using these strategies where appropriate:

Example

- **Use pictures**
- **Use smaller numbers**
- **Use bar models**
- **Use x for the unknown**
- **Use flow diagrams**
- **Use more bar models**
- **Use formulae**
- **Use arrow diagrams**
- **Use graphs**
- **Use problem-solving strategies and then write a sentence to 'explain' or 'show that …'.**

1   Alex, James and Emma are using sticks to race under a bridge. They drop three sticks at the same time on one side and see which appears first on the other side. They record who wins each time. After 30 races, Alex has won 17 times and James has won 5 times.
What is the estimated probability that Emma will win the next race?

2   **Exam-style question**

The diagram shows a small square inside a large square.

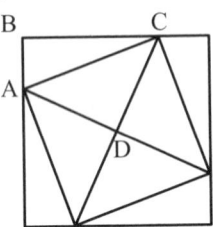

a   Copy the diagram. Shade the shapes that are congruent to triangle ACD. **(1 mark)**
b   The hypotenuse of triangle ACD is 8 cm in length. How long is the hypotenuse of each of the shapes that are congruent to triangle ACD?   **(1 mark)**

3   The height of Buckingham Palace is 24 metres. In a painting, Buckingham Palace has a height of 24 cm.
What is the scale factor of the enlargement?

4   **R** Triangles ABC and QRS are congruent.

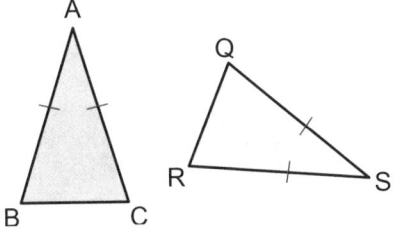

a   Which angle is the same as angle BAC?
b   Use the knowledge that angle ABC measures 69° to work out the size of the other five angles on the two triangles.

5   Scott is working on a construction project. He is calculating the pressure when a force of 50 N is applied to an area of 3.125 m². He uses the formula $P = \dfrac{F}{A}$.

a   What is the pressure?
b   What force would be applied if the pressure was 12 N/m²?

6   **R** Ryan makes a photocopy of this triangle.

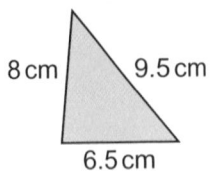

He wants the photocopy to be larger than the original.
a   Will Ryan's photocopy be the size he wants if he uses a scale factor of 0.5?
Explain why or why not.
b   Ryan's photocopied triangle has side lengths of 19 cm, 13 cm and 16 cm.
What is the scale factor of the enlargement?

7   **R** Mitchell draws a triangle similar to the one shown.

a   Explain what you know about Mitchell's triangle without seeing it.
b   Could one of the angles of Mitchell's triangle be 21°? Explain why or why not.
c   The longest side of Mitchell's triangle is 18 cm. What scale factor did he use?
d   Mitchell adds the three angles of his triangle. The first two he adds total 130°. What size is the third angle of Mitchell's triangle?

8   Raheem writes the vector $\begin{pmatrix} 3 \\ 5 \end{pmatrix}$.
Dani writes a vector that is parallel to Raheem's vector.
Write two possible vectors Dani could have written.

9   The vectors for a journey are $\overrightarrow{AB} \begin{pmatrix} 3 \\ 4 \end{pmatrix}$ and $\overrightarrow{BC} \begin{pmatrix} 4 \\ -2 \end{pmatrix}$.
What is the resultant vector $\overrightarrow{AC}$ for this journey?

10 Trish has been asked to factorise $x^2 - 11x + 24$.
What should Trish's answer be?

Q10 hint Think about the factors of 24.

20 MORE ALGEBRA

20.1 Graphs of cubic and reciprocal functions

1 a Copy and complete the table of values for $y = x^3$.

x	−3	−2	−1	0	1	2	3
y							

   b On graph paper, draw a coordinate grid with the x-axis from −3 to +3 and the y-axis from −30 to +30.

   c Plot the points from your table of values for $y = x^3$.

   d Join the points with a smooth curve. Label your graph with its equation.

2 a Use your graph from **Q1** to estimate

      i 1.7^3

      ii $\sqrt[3]{20}$

   b Use a calculator to work out

      i 1.7^3

      ii $\sqrt[3]{20}$

3 a Make a table of $y = -x^3$ for values of x from −3 to +3.

   b Plot the graph of $y = -x^3$.

4 a Copy and complete this table of values for $y = x^3 - 3$.

x	−3	−2	−1	0	1	2	3
x^3	−27					8	
−3	−3	−3	−3	−3	−3	−3	−3
y	−30					5	

   b Draw a pair of axes on graph paper. Plot the graph of $y = x^3 - 3$.

   c Draw a table of values for $y = x^3 + 5$.

   d Plot the graph of $y = x^3 + 5$ on your axes from part **b**.

   e What is the same about your two graphs? What is different?

5 Use your graph of $y = x^3 + 5$ from **Q4** to find the solution of $x^3 + 5 = 0$.

6

   a Copy and complete the table of values for $y = x^3 - x + 3$

x	−3	−2	−1	0	1	2	3
y	-27	−3	3	3	3	9	27

   (2 marks)

   b Copy the grid and draw the graph of $y = x^3 - x + 3$ from $x = -3$ to $x = 3$

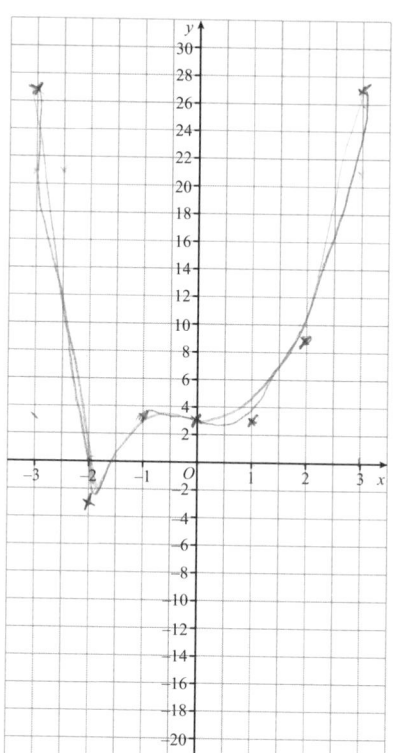

   (2 marks)

Exam hint

You might find it easier to draw your own table with a row for each part of the function.

x	
x^3	
$-x$	
$+3$	
y	

Complete the graph. Plot the points using crosses. Join them with a smooth curve. Always use a sharp pencil.

7   a   Copy and complete the table of values for $y = \dfrac{1}{x}$.

x	-4	-3	-2	-1	$-\frac{1}{2}$	$-\frac{1}{4}$	$\frac{1}{4}$	$\frac{1}{2}$	1	2	3	4
y	$-\frac{1}{4}$				-2			2	1			$\frac{1}{4}$

   b   Plot the points.
       Join the two parts with smooth curves.
   c   Label your graph $y = \dfrac{1}{x}$.

8   a   Sketch the graphs of $y = x^3$ and $y = \dfrac{1}{x}$.
       How can you remember which is which?
   b   How can you find the graph of $y = -x^3$ from your sketch in part **a**?

9   **R** Match each equation to a graph.
   a $y = x^3$ b $y = 5x$
   c $y = -3x$ d $y = -x^2$
   e $y = -x^3$ f $y = \dfrac{1}{x}$

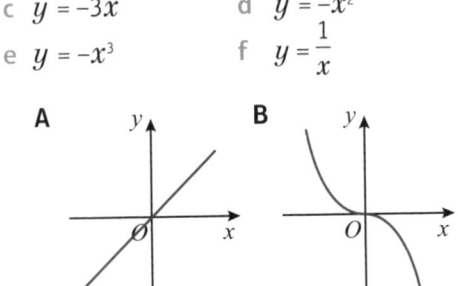

20.2 Non-linear graphs

1   **R** This 3D shape is made by removing a small cube from the centre of a larger cube.

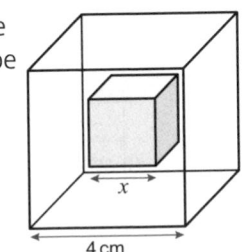

   a   Write an expression for the volume of the shape.

   b   Draw the graph of your expression in part **a**, for values of x from 0 to 4.
   c   Estimate the volume of the shape when $x = 3.1$ cm
   d   A shape like this has volume 20 cm³. What is the value of x?

   > Q1b hint Make a table of values first.

2   **P** This graph shows how long it takes different numbers of chefs to prepare meals in a restaurant.

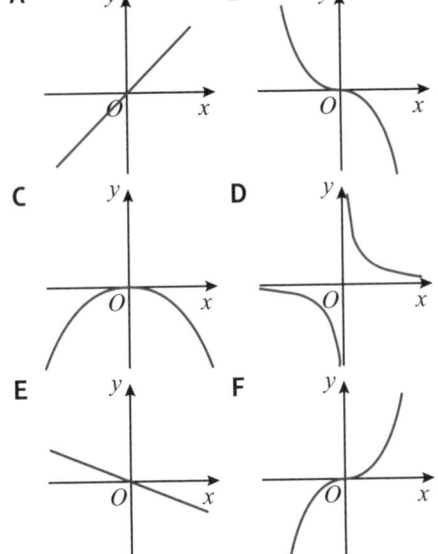

   a   Describe the relationship between the number of chefs and the number of minutes.
   b   Estimate how long it would take 3 chefs to prepare all the meals.
   c   The meals need to be prepared in 40 minutes. How many chefs are required?

3   **R** The graph shows the amount of gas in a helium balloon as it goes down.

   a   How much helium was in the balloon
       i   to start with   ii   after 3 hours?
   b   Estimate how many hours it took for
       i   half the helium in the balloon to escape
       ii   3500 cm³ to escape.
   c   Show that the balloon emptied less quickly as the volume of helium in it decreased.

4

Exam-style question

The table shows the current and resistance in five different circuits which each have a voltage of 60 V.

Current (amps)	1	2	3	4	5
Resistance (ohms)	60	30	20	15	12

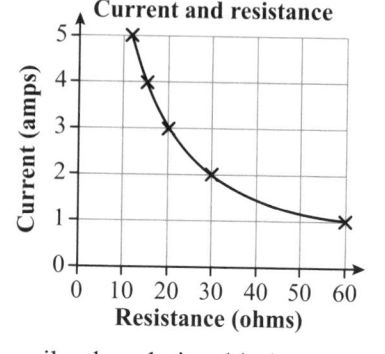

Current and resistance

a Describe the relationship between current and resistance. **(2 marks)**

b Estimate the current in a 60 V circuit with a resistance of 50 ohms. **(1 mark)**

c Amber makes a 60 V circuit and measures the current to be 2.5 amps. Estimate the resistance in her circuit. **(1 mark)**

5 **R** The graph shows the count rate against time for a radioactive material, sodium-24. The count rate measures the number of radioactive emissions per second.

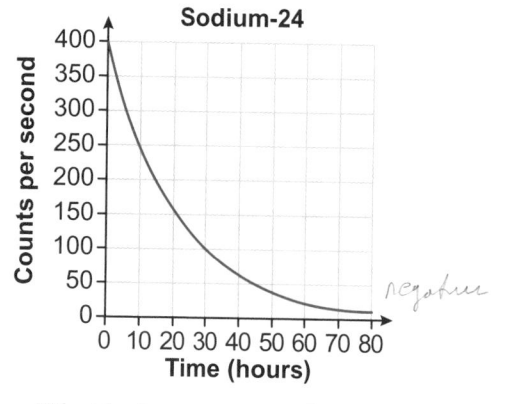

Sodium-24

a What is the count rate after 60 hours?

b How many hours does it take for the count rate to reduce to 50?

c The half-life of a radioactive material is the time it takes for the count rate to halve. What is the half-life of sodium-24?

6 The graph shows how the number of bacteria in a sample grows over a 4-hour period.

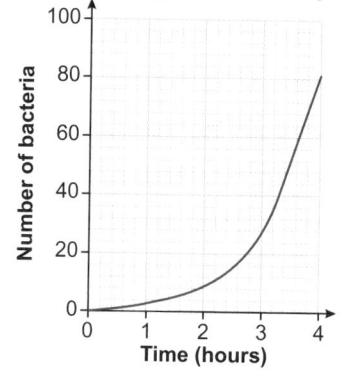

Bacteria growth in a sample

a Copy and complete the table to show the numbers of bacteria.

Time (hours)	0	1	2	3	4
Number of bacteria					

b Describe the sequence of the numbers of bacteria.

c Use the graph to estimate the number of bacteria after $2\frac{1}{2}$ hours.

d Estimate when the number of bacteria reaches 70.

7 **P** The graph shows the value of an investment over a 5-year period.

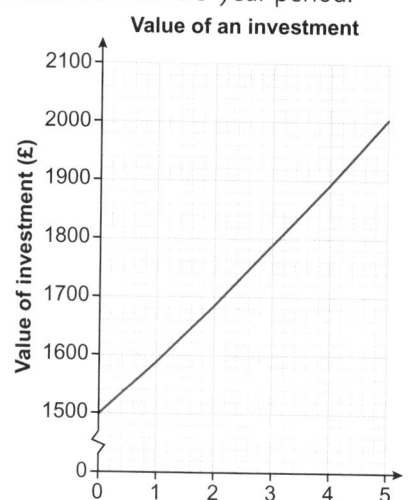

Value of an investment

a What was the initial value of the investment?

b Estimate the value of the investment after 5 years.

c By how much did the value increase in the first year?

d The rate of interest remained the same for the five years.

Work out the percentage interest rate.

1 a Draw the graph of $x + 2y = 10$.
   b Draw the graph of $y = 3$ on the same axes.
   c Write the solution to the simultaneous equations $x + 2y = 10$ and $y = 3$.

2 Draw graphs to solve these simultaneous equations.
   $3x + y = 10$ and $x - 2y = 1$

3 Solve these simultaneous equations graphically.
   $x - y = -4$ and $2x + 5y = 6$

4 Use these graphs to solve the pairs of simultaneous equations.
   a $x + y = 6$
      $3x - y = 5$
   b $x + y = 6$
      $x + 3y = 10$
   c $3x - y = 5$
      $x + 3y = 13$

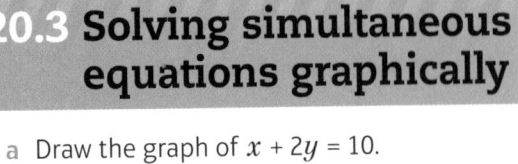

5 **Exam-style question**

   The graph of $2x + 3y = 3$ is shown on this grid.

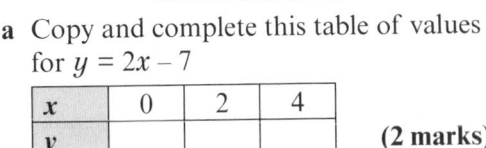

   a Copy and complete this table of values for $y = 2x - 7$

x	0	2	4
y			

   (2 marks)

   b Copy the graph grid and the line $2x + 3y = 3$
      Draw the line $y = 2x - 7$ on the grid.
      (2 marks)

   c Use your graphs to solve the simultaneous equations:
      $2x + 3y = 3$
      $y = 2x - 7$
      (2 marks)

6 **R** By drawing their graphs, show that the two equations
   $2x + 2y = 7$ and $2x + 2y = -2$
   have no solution.

7 **R** Adam is older than Beth. The sum of Adam's and Beth's ages is 21. The difference in their ages is 5 years.

   Example

   a Write an equation for the sum of their ages.
   b Write an equation for the difference in their ages.
   c Draw the graph of each equation.
   d What are Adam's and Beth's ages?

8 **P** A teacher is planning a trip to the theatre. The costs for two different classes are
   Class 1: 2 adults and 11 children cost £101
   Class 2: 3 adults and 20 children cost £176.
   a What do x and y stand for in this equation for Class 1? $2x + 11y = 101$
   b Write an equation for Class 2.
   c Draw graphs to find the cost of an adult ticket and the cost of a child ticket.

9 **P** Zoe buys 2 big packets and 1 small packet of biscuits, which contain 76 biscuits in total. Ross buys 1 big packet and 2 small packets of biscuits, which contain 56 biscuits in total.
   a Write equations for the biscuits that Zoe and Ross each buy.
   b Solve your simultaneous equations graphically to work out the number of biscuits in a big packet and the number of biscuits in a small packet.

20.4 Solving simultaneous equations algebraically

1 Here are two simultaneous equations:
   $x + 2y = 10$ and $y = 3$.
   a Substitute $y = 3$ into the equation $x + 2y = 10$
   b Solve your equation in x from part **a**.
   c Write the solution to the simultaneous equations.
   d In Lesson 20.3 **Q1**, you drew graphs to solve these equations.
      Are your solutions the same?

2 Solve these simultaneous equations algebraically:
   $x + y = 6$ and $x + 3y = 10$
   In Lesson 20.3 **Q4b**, you solved these equations using graphs. Are your solutions the same?

Example

3 Solve these pairs of simultaneous equations.
   a $x + 3y = 9$ and $-x + 5y = 7$
   b $3x + y = 11$ and $5x + y = 15$
   c $-x + 2y = 5$ and $-x + 6y = 11$

4 **R** The sum of two numbers is 22 and their difference is 8.
   Write and solve a pair of simultaneous equations to find the two numbers.

5 Solve these simultaneous equations.
   $3x + y = 13$ (1)
   $2x - 2y = 6$ (2)

   Q5 hint Multiply the first equation by 2 and label this equation (3). Now solve equations (2) and (3).

6 **Exam-style question**

   Solve the simultaneous equations
   $5x + 2y = 18$
   $7x + 6y = 38$ **(3 marks)**

7 Solve these simultaneous equations.
   $2x + 7y = 19$ (1)
   $5x + 3y = 4$ (2)
   a First multiply equation (1) by 5. Label this equation (3).
   b Now multiply equation (2) by 2. Label this equation (4).
   c Solve the simultaneous equations (3) and (4).

8 Repeat **Q7**, but this time
   a Multiply equation (1) by 3. Label this equation (3).
   b Multiply equation (2) by 7. Label this equation (4).
   c Solve the simultaneous equations (3) and (4).

9 **P** Yasmin is paid £65 for 8 hours' work plus 1 hour's overtime.
   For 11 hours' work plus 3 hours' overtime she is paid £104.
   How much is she paid for
   a 1 hour's work
   b 1 hour's overtime?

10 Follow the steps to find the equation of the line that passes through the points A(3, 7) and B(5, 13).
   a Copy and complete these two equations for the line.
      Use the x and y values from each coordinate pair in $y = mx + c$
      At point A: $\Box = 3m + c$
      At point B: $\Box = 5m + c$
   b Solve the two simultaneous equations to find the values of m and c.
   c Use your values of m and c to write the equation of the line.

11 **P** Find the equation of the line that passes through
   a C(2, 9) and D(4, 13)
   b E(1, 1) and F(−2, 4)

20.5 Rearranging formulae

1 What is the subject of each formula?
   a $P = 4e$ b $r = \dfrac{d}{2}$
   c $F = ma$ d $W = Fd$

2 Copy and complete to make x the subject of each formula.
   a $K = 3x$ so $\dfrac{K}{\Box} = \Box$
   b $N = \dfrac{x}{5}$ so $5N = \Box$

3 Make the letter in brackets the subject of each formula.
   a $P = 4e$ [e]

   b $F = ma$ [a]

   c $I = \dfrac{Q}{t}$ [t]

   d $P = \dfrac{E}{t}$ [t]

   e $V = \dfrac{W}{Q}$ [W]

   f $s = \dfrac{d}{t}$ [t]

   Q3d hint Rearrange $P = \dfrac{E}{t}$ to $pt = E$ and then $t = \dfrac{\Box}{\Box}$

4 This is a density–mass–volume triangle used in science to help remember formulae.

From the triangle you can see that

$D = \dfrac{M}{V}$, $M = DV$ and $V = \dfrac{M}{D}$.

Now rearrange $D = \dfrac{M}{V}$ to make

a M the subject of the formula

b V the subject of the formula.

Do you get the same formulae?

Which is easier to remember – the triangle

or how to change the subject of $D = \dfrac{M}{V}$?

5 The table shows the average speed and distance for three different journeys.

speed $(s) = \dfrac{\text{distance } (d)}{\text{time } (t)}$

Work out the time taken for each journey in hours, correct to 3 d.p.

Transport	Speed (km/h)	Distance (km)
bike	19	45
car	97	212
helicopter	135	185

6 Change the subject of each formula to the letter in brackets.

Example

a $y = mx + c$ [x]

b $p = qr - t$ [q]

c $d = a - bc$ [b]

d $x = u + vw$ [w]

7 a Rearrange the equation of the line $6x + 3y = 10$ into the form $y = mx + c$

b What are the gradient and the y-intercept of the line?

8 **P** Which of these lines pass through the point $(0, 4)$?

a $y = 3x + 4$ b $2y + 2x = 4$

c $2y - 4x = -16$ d $3y - x = 12$

e $x = 2y - 8$

9 **P** Tariq wants to draw three squares with areas 10 cm², 30 cm² and 40 cm².

How long should he draw the sides of each square?

Round your answers to a sensible degree of accuracy.

10 Make the letter in brackets the subject of each formula.

a $C = prl$ [l]   b $A = pdh$ [h]

c $A = pdh$ [d]   d $l = \dfrac{prs}{180}$ [s]

11 Make the letter in brackets the subject of each formula.

a $K = \dfrac{ABC}{2}$ [A]

b $r = \dfrac{2A}{l}$ [A]

c $P = 2(a + b)$ [a]

d $V = \dfrac{abh}{2}$ [h]

e $N = p(p + q)$ [q]

f $A = \dfrac{p(a + b)}{2}$ [b]

g $r = \dfrac{C}{2p}$ [p]

h $F = 2n(b - c)$ [b]

12 a Copy and complete to make x the subject of $y = \dfrac{x - 2}{n} + 5$

$y - \square = \dfrac{x - 2}{n}$

$\square (y - \square) = x - 2$

$\square (y - \square) + \square = x$

b Make x the subject.

i $p = \dfrac{x + 1}{q} - 4$

ii $k = \dfrac{x - 4}{j} + m$

iii $3x - a = 2x - 5$

iv $4x + t = 7x - r$

13 Make the letter in brackets the subject.

Example

a $a = b^2$ [b]

b $y = 3x^2$ [x]

c $r = \frac{1}{2}t^2$ [t]   d $A = \dfrac{p^2}{16}$ [p]

e $P = \sqrt{q}$ [q]   f $C = \sqrt{5d}$ [d]

g $M = \sqrt{j + k}$ [j]   h $E = \frac{1}{2}mv^2$ [v]

i $A = 2\pi r^2$ [r]   j $a = \dfrac{v^2}{r}$ [v]

k $E = mc^2$ [c]   l $c^2 = a^2 + b^2$ [b]

14 **P** Each of these cylinders has a volume of 100 cm³.
Work out the missing dimensions.

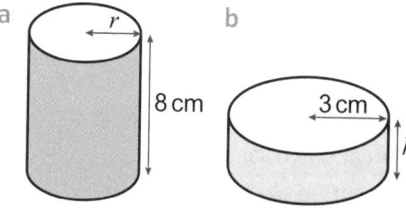

a

b

15
> **Exam-style question**
>
> Make x the subject of the formula
> $3(x + y) = 5x - 8$ **(3 marks)**

16 These formulae connect final velocity (v), initial velocity (u), acceleration (a), distance (s) and time (t).

a Make v the subject of $s = vt - \frac{1}{2}at^2$

b Make t the subject of $s = \dfrac{(u + v)t}{2}$

c Make u the subject of $s = \dfrac{(u + v)t}{2}$

20.6 Proof

1 **R** a Solve $2x + 7 = 19$

b Is $2x + 7 = 19$ true for all values of x?

c Is $2x + 7 = 19$ an equation or an identity?

2 **R** Decide if these are expressions, formulae, equations or identities.

a $x^2 + x + 1$

b $5x - 2 = 33$

c $E = mgh$

d $3(x + 2) = 3x + 6$

3 Show that

a $x(x + 5) \equiv x^2 + 5x$

b $(x - 1)^2 + 2x \equiv x^2 + 1$

c $x^2 + 5x + 10 \equiv (x + 3)^2 - x + 1$

d $3x^3 - 2x^2 \equiv x^2(3x - 2)$

e $x^2 - 25 \equiv (x + 5)(x - 5)$

f $(x + y)(x - y) \equiv x^2 - y^2$

Example

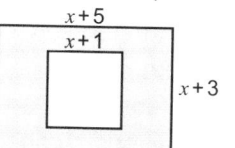

4 Follow the steps to show that
$(x + 4)(x + 6) + 1 \equiv (x + 5)^2$

a Multiply out the brackets on the LHS.

b Simplify the LHS by collecting like terms.

c Multiply out the brackets on the RHS.

d Show that the LHS and RHS are the same.

5 A rectangular card with length $x + 5$ and width $x + 3$ has a smaller square cut out of the middle. The square has sides of length $x + 1$.

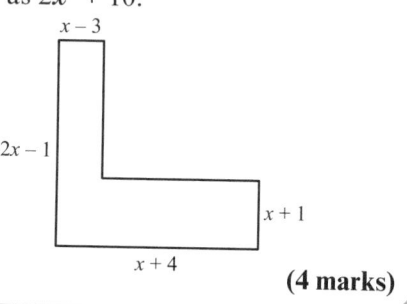

a Write an expression for the total area of the rectangle before the middle was cut out.

b Write an expression for the area of the square that has been cut out.

c Show that the area of the remaining card is $6x + 14$.

6
> **Exam-style question**
>
> Show that the area of this shape can be written as $2x^2 + 10$.
>
>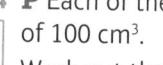
>
>   **(4 marks)**

7 a Write any four consecutive integers.
Make the first one equal to n.
Write expressions for the other three numbers.

b Write a different set of four consecutive numbers.
Make the third one equal to n and write expressions for the other three numbers.

8 The median of a set of three consecutive integers is n.

a Write expressions for the three integers.

b Show that the range of the three integers is 2.

c Show that the mean of the three integers is n.

9 Follow these steps to show that the sum of any five consecutive integers is a multiple of 5.

a Write expressions for the five integers as
$n,$ ___, ___, ___, ___ .

b Work out the sum of these integers.

c Simplify the sum as much as possible, then factorise it.

d Copy and complete: The sum of five consecutive numbers is _____, which is ____ × 5, which is a multiple of 5.

10 Show that the sum of any four consecutive numbers is not a multiple of 4.

11 **R** a Work out the first five terms of the sequence with general (nth) term
   i $2n$ ii $2n + 1$.
   b What is the general term for
   i an even number   ii an odd number?

12 **R** Copy and complete this proof to show that the sum of an even number and an odd number is odd.
An even number is $2m$ and an odd number is $2n + 1$.
$2m + 2n + 1 = \square(\square + \square) + 1$
$\square(\square + \square)$ is a multiple of $\square$, so it is an even number.
Therefore $\square(\square + \square) + 1$ is an odd number.

13 **R** Show that the product of two even numbers is even.

14 **R** Given that $4(x + n) = 3x + 10$, where n is an integer, show that x must be an even number.

20 Problem-solving

Solve problems using these strategies where appropriate:
- **Use pictures**
- **Use smaller numbers**
- **Use bar models**
- **Use x for the unknown**
- **Use flow diagrams**
- **Use more bar models**
- **Use formulae**
- **Use arrow diagrams**
- **Use graphs**
- **Use problem-solving strategies and then write a sentence to 'explain' or 'show that …'.**

1 a What can you do to find the values for $y = 2x + 3$ before you plot the graph?
   b What does the graph $y = 2x + 3$ look like?

2 **Exam-style question**
   a Complete the table of values for $y = 2x - 1$

x	-2	-1	0	1	2	3
y	-5	-3		1		5

   (2 marks)
   b Draw the graph of $y = 2x - 1$ for values of x from -2 to 3. **(2 marks)**

3 **R** I took a number from a set of number cards and doubled it. I then added the total to the next number I took from the cards. My sum was 17. The second number was 2 more than the first.
Work out the two numbers.

4 **R** The lengths of two of the sides of a rectangular patch for a quilt are $x + 3$ and $x + 4$.
   a Work out an expression for the area of the patch. Simplify your answer.
   b Eight patches are used to create a pattern on the quilt. Write an expression for the total area of the pattern. Simplify your answer.

5 **R** Philip says that
$(x + 1)^2 - 7(x + 2) \equiv (x + 3)^2 - 11(x + 2)$.
Kyle says that it can't be true.
Who is correct?

6 Five years ago, Carl bought £50 of shares from the company he works for. In the first year the value of the shares increased to £57.50. In the fourth year the shares were worth £87.45 and now the shares are valued at £100.57.
   a What was the value of the shares in the third year?
   b How much will the shares be worth after
   i 6 years
   ii 10 years
   if the increase continues at the same rate?

7 Matthew translates a shape by the column vector $\begin{pmatrix} 5 \\ 2 \end{pmatrix}$ and then by the column vector $\begin{pmatrix} -2 \\ 3 \end{pmatrix}$.
Find the column vector of Matthew's work as a single translation.

8 Jayne is measuring the height of trees. She has an instrument that tells her the angle from the ground to the top of the tree. Jayne stood 15 feet away from the tree and measured an angle of 61° from the ground to the top of the tree.
How many feet tall is the tree?
Round your answer to 1 d.p.

9 Make r the subject of the formula $b = \dfrac{4s + r}{t}$

10 Show clearly that the volume of this cuboid is $2x^3 + 6x^2 + 4x$.

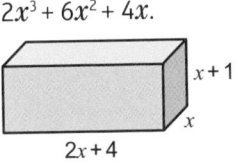

$x + 1$
x
$2x + 4$